Application of Superabsorbent Polymers (SAP) in Concrete Construction

RILEM STATE-OF-THE-ART REPORTS

Volume 2

RILEM, The International Union of Laboratories and Experts in Construction Materials, Systems and Structures, founded in 1947, is a non-governmental scientific association whose goal is to contribute to progress in the construction sciences, techniques and industries, essentially by means of the communication it fosters between research and practice. RILEM's focus is on construction materials and their use in building and civil engineering structures, covering all phases of the building process from manufacture to use and recycling of materials. More information on RILEM and its previous publications can be found on www.RILEM.net.

The RILEM State-of-the-Art Reports (STAR) are produced by the Technical Committees. They represent one of the most important outputs that RILEM generates – high level scientific and engineering reports that provide cutting edge knowledge in a given field. The work of the TCs is one of RILEM's key functions.

Members of a TC are experts in their field and give their time freely to share their expertise. As a result, the broader scientific community benefits greatly from RILEM's activities.

RILEM's stated objective is to disseminate this information as widely as possible to the scientific community. RILEM therefore considers the STAR reports of its TCs as of highest importance, and encourages their publication whenever possible.

The information in this and similar reports is mostly pre-normative in the sense that it provides the underlying scientific fundamentals on which standards and codes of practice are based. Without such a solid scientific basis, construction practice will be less than efficient or economical.

It is RILEM's hope that this information will be of wide use to the scientific community.

For further volumes:
http://www.springer.com/series/8780

Viktor Mechtcherine • Hans-Wolf Reinhardt
Editors

Application of Superabsorbent Polymers (SAP) in Concrete Construction

State of the Art Report Prepared by Technical Committee 225-SAP

Springer

Editors
Viktor Mechtcherine
Faculty of Civil Engineering
Institute of Construction Materials
Technische Universität Dresden
01062 Dresden
Germany
mechtcherine@tu-dresden.de

Hans-Wolf Reinhardt
Faculty of Construction
and Environmental Engineering Sciences
Department of Construction Materials
University of Stuttgart
70569 Stuttgart
Germany
reinhardt@iwb.uni-stuttgart.de

ISBN 978-94-007-2732-8 e-ISBN 978-94-007-2733-5
DOI 10.1007/978-94-007-2733-5
Springer Dordrecht Heidelberg London New York

Library of Congress Control Number: 2011944021

Printed on acid-free paper

Springer is part of Springer Science+Business Media (www.springer.com)

Foreword

The increasing interest in the use of SAP as a concrete additive and the need for intensive scientific exchange among research groups led in 2007 to the initiation of the RILEM Technical Committee 225-SAP "Application of Superabsorbent Polymers in Concrete Construction". This committee brings together from different countries around the world recognized researchers who are presently investigating the mechanisms of SAP action in concrete materials and the possibilities and limitations of using SAP as a set of solutions to various problems encountered by practitioners in the field. The committee had meetings in Delft, The Netherlands (May 2008), Ise-Shima, Japan (September 2008), Dresden, Germany (March 2009), Haifa, Israel (September 2009), Aachen, Germany (September 2010) and Stuttgart, Germany (July 2011).

The committee's objective is to coordinate research efforts and compile the results of studies with respect to the effects of SAP addition on the properties of concrete in its fresh and hardened states. This State-of-the-Art Report is the main product of the committee's work. It summarizes the available information and knowledge in the area and provides as well a solid basis and a good reference for further research. Also, it is to serve as starting point for further activity of RILEM TC 225-SAP, including a series of round-robin tests and the development of practical recommendations for utilizing SAP in concrete construction.

Because this report is to provide a comprehensive yet easy-to-follow overview of different aspects of the use of SAP as a concrete additive, it was decided to subdivide the book into ten chapters, each covering particular area of interest. The presentational sequence of the topics was chosen according to principles "from fundamentals to applications" and "from fresh to hardened concrete".

Each chapter had a chapter coordinator, who was also the first author. Additionally, the committee members who contributed significantly to the shaping and writing of the corresponding chapter are named as co-authors. The chapters' contents were discussed comprehensively and approved in committee meetings and by email correspondence among the members. W. Brameshuber, D. Cusson, K. Kovler,

V. Mechtcherine and H.-W. Reinhardt reviewed the individual chapters for their content. The last two persons carried out the editorial work. J. Weiss proofread the entire manuscript with respect to its editorial correctness.

Dresden Viktor Mechtcherine

Contents

Acknowledgement

Prepared by: RILEM TC 225-SAP

Chairman: MECHTCHERINE, Viktor, Germany

Secretary: REINHARDT, Hans-Wolf, Germany

Members: ASSMANN, Alexander, Germany
BAROGHEL-BOUNY, Véronique, France
BETTENCOURT RIBEIRO, António, Portugal
BRAMESHUBER, Wolfgang, Germany
CUSSON, Daniel, Canada
DE BELIE, Nele, Belgium
DE LA VARGA, Igor, USA
DUDZIAK, Lukasz, Germany
DURÁN-HERRERA, Alejandro, Mexico
ESTEVES, Luis P., Portugal
FONSECA DA SILVA, Eugênia, Brazil
FRIEDEMANN, Karen, Germany
FRIEDRICH, Stefan, Germany
HASHOLT, Marianne, Denmark
ICHIMIYA, Kazuo, Japan
IGARASHI, Shin-ichi, Japan
JENSEN, Ole Mejlhede, Denmark
KLEMM, Agnieszka J., United Kingdom
KOENDERS, Eddie A. B., The Netherlands
KOVLER, Konstantin, Israel
LAUSTEN, Sara, Denmark
LURA, Pietro, Switzerland
MIHASHI, Hirozo, Japan
MÖNNIG, Sven, Germany
RIEDER, Klaus-Alexander, Germany
STALLMACH, Frank, Germany

TOLEDO FILHO, Romildo D., Brazil
VAN BREUGEL, Klaas, The Netherlands
WEISS, Jason, USA
WYRZYKOWSKI, Mateusz, Switzerland
YE, Guang, The Netherlands

Corresponding author: Prof. Dr.-Ing. Viktor Mechtcherine
Institute of Construction Materials
Faculty of Civil Engineering
Technische Universität Dresden
01062 Dresden
Germany
E-mail: mechtcherine@tu-dresden.de
Phone: +49 351 463 35 920
Fax: +49 351 463 37 268

Chapter 1
Introduction

Viktor Mechtcherine

Abstract This chapter is a short introduction to the State-of-the-Art Report on the application of Superabsorbent Polymers (SAP) in concrete construction. It describes the general role of chemical additives in concrete technology, outlines the possible functions of SAP in cement-based materials and defines the prospective main areas of this new additive's use. Additionally, the concept and structure of the book are briefly explained.

1.1 SAP as a New Concrete Additive

In the last few decades great advances in concrete technology have arisen, to a large extent out of the development and use of new chemical additives which although added to concrete in very small quantities can dramatically improve crucial properties of concrete in its fresh and/or hardened state. One prominent example is the use of modern superplasticizers. When superplasticizers are used with other appropriate ingredients they enable the development of new types of concrete such as Self-Compacting Concrete or Ultra-High-Performance Concrete. However, despite these considerable advances and the already broad palette of existing concrete additives, there is a great need for further progress.

One of the key considerations in concrete technology is gaining control of the water. On the one hand, some amount of water is needed to hydrate the cement and to achieve the required rheological properties of the various concrete materials in their mixing, transporting, placing, and compacting. On the other hand, with increasing free water content the danger of segregation and bleeding of fresh

V. Mechtcherine (✉)
Institute of Construction Materials, Technische Universität Dresden, Germany
e-mail: mechtcherine@tu-dresden.de

V. Mechtcherine and H.-W. Reinhardt (eds.), *Application of Superabsorbent Polymers (SAP) in Concrete Construction*, RILEM State of the Art Reports 2,
DOI 10.1007/978-94-007-2733-5_1, © RILEM 2012

concrete increases. Furthermore, it leads to the increased porosity of the hardened concrete and accordingly to considerable reduction of its mechanical performance, reduction in durability, reduced resistance to permeation, and increased shrinkage and creep deformations. Water-reducing additives like the superplasticizers already mentioned make it possible to achieve good workability of the fresh concrete and correspondingly a dense concrete microstructure in the hardened state. Stabilizing additives such as, for a single example, methylcellulose first of all affect the availability of free mixing water and reduce therewith the tendency of fresh mixes toward bleeding and segregation.

The introduction of SAP as a new component for the production of concrete materials makes available a number of new possibilities with respect to water control and, as a result, to the control over the rheological properties of fresh concrete, in addition to purposeful water absorption and/or water release in either fresh or hardened concrete. A well controlled uptake and release of water can be fostered by the specific design of SAP materials adapted to particular practical needs. As examples of this, the internal curing of High-Performance Concrete (see [1, 2] and Chapter 7 of this report) and the inducing of an abrupt change in rheological behavior during shotcreting (see [3] and Chapter 10) might be given here, but the potential for innovation is far wider.

Another persistent problem in modern concrete technology relates to the creation in concrete of advantageous pore systems which could improve its durability, especially in terms of freeze-thaw resistance. Contemporary air-entrainment agents are widely used in achieving such high freeze-thaw resistance. However, in practice this technique often falls far short of its goal. The entrained air voids are frequently not stable enough to sustain transport, compacting, or in some instances specific methods of application such as spraying. The upshot is that there is currently strong demand for more robust solutions, as one of which SAP could be regarded already as a good alternative. Pore systems built up as a result of SAP addition seem to remain stable regardless of the consistency of the concrete, the addition of superplasticizer, or the method of placement and compacting. And the freeze-thaw resistance of concrete with SAP added is comparable with that of well working, air-entrained concrete (see [4] and Chapter 9).

Further applications have been proposed, some as vague ideas as well as some supported by preliminary investigations. An example of such suggestions is the utilization of SAP as micro-reservoirs for chemical substances which would be released under specific conditions such as temperature, changes of the chemical composition of pore solutions, passage of time, etc. [1]. Also, the use of SAP as a multifunctional additive has been recently demonstrated. The approach was to improve several properties of Strain-hardening Cement-based Composite (SHCC) simultaneously, using to advantage various aspects of action of SAP. The SAP particles act as micro-defects which trigger formation of multiple cracks when SHCC is subjected to high tensile loading, thereby increasing its ductility. At the same time SAP acts as an additive for increasing freeze-thaw resistance of the composite and as internal curing agent [5].

1.2 RILEM TC 225-SAP and Purpose of this Report

The initial ideas for the use of SAP as additives for construction applications were expressed immediately following the development of this new group of polymers. The first patents were written by DOW and Hoechst, dealing with dry mortars containing superabsorbents (see Chapter 3). However, there is very little knowledge of the spectrum of SAP applications in concrete construction, even though there are many indications that such applications already exist. Many experts working in the field of concrete technology have never heard of the possibility of using SAP as an additive, and no product indicated as "SAP" appears to have been produced and offered specifically for application as a concrete additive. Nevertheless, several examples are known where SAP are used as main components in an additive, but without indication of this in the technical documentation.

Due to the increased research activity at various independent research institutions, this "mysterious" situation might well change for the better in the near future. The State-of-the-Art Report of the RILEM TC 196-ICC "Internal Curing of Concrete" [6] describes the positive effects of the use of SAP in High-Strength Concrete as an efficient internal curing agent. Through the use of SAP, self-desiccation and the resulting autogenous shrinkage of such concretes could be reduced or even completely avoided by providing internal water reservoirs. Because of their very high water absorption capacity, SAP are more effective in mitigating autogenous shrinkage than other materials. A pilot project followed, providing evidence that internal curing based on use of SAP can be applied to High-Performance Concretes on a large scale [2].

The work of the RILEM TC 196-ICC has triggered much broader research on the topic of SAP. A number of researchers investigated the effects of adding SAP on concrete's rheological behaviour, shrinkage, strength, durability, and other properties (see Chapters 5, 7, 8, and 9, respectively). Furthermore, considerable efforts have gone into the investigation of mechanisms of SAP action on various concrete properties such as the kinetics of water absorption and desorption through SAP as well as on changes in concrete microstructure (see Chapters 4 and 6). The number of scientific publications on subjects related to the application of SAP in concrete construction has increased dramatically over the last few years. However, the findings of these investigations are to an extent the subject of some controversy, which can to a great extent be ascribed to the sensitivity of the results to the type of SAP in use, the amount of additional mix water, and other variations in the concrete compositions under examination.

The increasing interest in the use of SAP as a concrete additive and the need for intensive scientific exchange among research groups led in 2007 to the initiation of the RILEM Technical Committee 225-SAP "Application of Superabsorbent Polymers in Concrete Construction". This committee brings together from different countries around the world recognized researchers who are presently investigating the mechanisms of SAP action in concrete materials and the possibilities and limitations of using SAP as a set of solutions to various problems encountered by practitioners in the field. The committee's objective is to coordinate research efforts and compile the results of studies with respect to the effects of SAP addition on the properties of concrete in its fresh and hardened states. This State-of-the-Art Report is the main product of the committee's work. It summarizes the available information and knowledge in the area and provides as well a solid basis and a good reference for further research. Also, it is to serve as starting point for further activity of RILEM TC 225-SAP, including a series of round-robin tests and the development of practical recommendations for utilizing SAP in concrete construction.

1.3 Concept and Structure of the Report

Because this report is to provide a comprehensive yet easy-to-follow overview of different aspects of the use of SAP as a concrete additive, it was decided to subdivide the book into ten chapters, each covering particular area of interest. The presentational sequence of the topics was chosen according to principles "from fundamentals to applications" and "from fresh to hardened concrete".

Following the introduction in Chapter 1, Chapter 2 gives the definitions of terms used in the report.

Chapter 3 is dedicated to the superabsorbent polymers themselves. Chemical composition, production techniques, relevant properties and their characterization are described. Furthermore, some hints are given on specific requirements with regard to the choice of SAP and the manner of adding SAP to concrete mixtures.

Chapter 4 deals with the kinetics of water migration, i.e., the absorption of pore solution during mixing and its release when the cement paste self-desiccates or is exposed to drying. Knowledge of the kinetics of water migration into and out of the SAP is essential to understand and optimize the internal curing of concrete as well as for other prospective practical applications of SAP. For specific cases where experimental results on SAP are missing in the literature, results obtained with other comparable agents are presented and the applicability to SAP is discussed. A final section of the chapter is dedicated to modeling the internal curing of concrete by means of SAP and especially to the modeling of water migration to and from the SAP.

Available information on the effects of SAP on workability is presented and discussed in Chapter 5. The chapter concludes that serious research efforts are still required to understand the influence of SAP addition on rheological behavior of concrete and mortar in their fresh states.

Chapter 6 focuses on the hardening process of binder paste and microstructure development. It shows that the addition of SAP changes the hydration process and the development of the microstructure in concrete. The degree of hydration of cement in composites containing SAP as well as the influence of SAP on the development of total porosity, pore size distribution, morphology and pore connectivity of bulk cement paste, and the interface transition zone between cement paste and SAP are discussed. Furthermore, the distribution of SAP particles in the mixture is addressed.

Chapter 7 deals with one of the main potential applications of SAP in concrete construction, i.e., mitigation of the autogenous shrinkage of concrete. In particular, the effects of the use of SAP as an additive for internal curing in cementitious materials with low w/c and little permeable microstructure are presented and discussed. Since the addition of SAP, often in conjunction with extra water, influences not only autogenous shrinkage but also other types of volumetric changes, this subject is addressed as well. Furthermore, development of stresses due to restraint of autogenous shrinkage is considered for concretes with and without internal curing.

Chapter 8 addresses the effect of adding SAP (with or without extra water) to concrete mixtures on the mechanical properties of the hardened concrete, such as its compressive and tensile strength as well as on its elastic properties.

In Chapter 9 various aspects of durability influenced by SAP are presented and discussed. In particular it contains findings and conclusions with regard to water and oxygen permeability, freeze-thaw resistance, and chloride migration.

The last chapter of the report, Chapter 10, is dedicated to practical applications of superabsorbent polymers in concrete and other building materials. It presents existing and projected opportunities for the use SAP in many different functions to improve the performance and durability of the built environment. Two case studies are also presented.

Each chapter had a chapter coordinator, who was also the main author. Additionally, the committee members who contributed significantly to the shaping and writing of the corresponding chapter are named as co-authors. The chapters' contents were discussed comprehensively and approved in committee meetings and by email correspondence among the members. W. Brameshuber, D. Cusson, K. Kovler, V. Mechtcherine and H.-W. Reinhardt reviewed the individual chapters for their content. The last two persons carried out the editorial work. J. Weiss proofread the entire manuscript with respect to its editorial correctness.

References

[1] Jensen OM, Hansen PF (2001) Water-Entrained Cement-Based Materials: I. Principle and Theoretical Background. Cem Concr Res 31(4):647–654

[2] Mechtcherine V, Dudziak L, Schulze J, Stähr H (2006) Internal curing by Superabsorbent Polymers – Effects on material properties of self-compacting fibre-reinforced high performance concrete. In: Jensen OM, Lura P, Kovler K (eds) Proceedings of international RILEM conference on Volume Changes of Hardening Concrete: Testing and Mitigation, 20-23 August 2006 (Technical University of Denmark, Lyngby, Denmark), pp 87–96

[3] Jensen OM (2008) Use of superabsorbent polymers in construction materials. In: Sun Wei et al. (eds) Proceedings of the 1st international conference on Microstructure Related Durability of Cementitious Composites, 13-15 October 2008 (Nanjing, China), pp 757–764

[4] Reinhardt H-W, Assmann A, Moennig S (2008) Superabsorbent polymers (SAP) – an admixture to increase the durability of concrete. In: Sun W et al. (eds) Proceedings of the 1st international conference on Microstructure Related Durability of Cementitious Composites, 13-15 October 2008 (Nanjing, China), pp 313–322

[5] Bruedern A-E, Mechtcherine V (2010) Multifunctional use of SAP in Strain-hardening Cement-based Composites. In: Jensen OM, Hasholt MT, Laustsen S (eds) Proceedings of international RILEM conference on Use of Superabsorbent Polymers and Other New Additives in Concrete, 15-18 August 2010 (Technical University of Denmark, Lyngby, Denmark), pp 11–22

[6] RILEM Report 41 (2007) State of the Art Report of RILEM Technical Committee Internal Curing of Concrete. Kovler K, Jensen OM (eds), RILEM Publications S.A.R.L., Bagneux, France, 141 pp.

Chapter 2
Terminology

Hans-Wolf Reinhardt, Daniel Cusson, and Viktor Mechtcherine

Abstract This chapter introduces the terminology used in the following chapters and provides the notation of abbreviations.

2.1 Terminology

Acrylic acid-co-acrylamide - hydrogel, owing to the existence of hydrophilic COOH and NH_2 groups, which has the capacity to absorb large amounts of water.

Addition - fine-grained inorganic material added to cement with the aim to improve some specific cementing properties. Includes two types [1]: inert or nearly inert additions (type I) and pozzolanic or latent hydraulic additions (type II). Quantities are usually ranging from 5% to 50% per unit mass of powder.

Admixture - material added to the concrete mixture in small quantities at time of mixing to modify the properties of fresh and/or hardened concrete. Quantities are usually ranging from 0.2% to 4% per unit mass of powder.

Air-entrainment - intentional introduction of air uniformly distributed in very small bubbles/voids into concrete primarily to improve frost resistance. Volume of entrained air usually varies between 3% and 7% of the total volume of concrete.

H.-W. Reinhardt (✉)
Department of Construction Materials, University of Stuttgart, Germany
e-mail: reinhardt@iwb.uni-stuttgart.de

D. Cusson
Institute for Research in Construction, National Research Council Canada, Canada

V. Mechtcherine
Institute of Construction Materials, Technische Universität Dresden, Germany

V. Mechtcherine and H.-W. Reinhardt (eds.), *Application of Superabsorbent Polymers (SAP) in Concrete Construction*, RILEM State of the Art Reports 2,
DOI 10.1007/978-94-007-2733-5_2, © RILEM 2012

Autogenous shrinkage - external, macroscopical (bulk) dimensional reduction (volume or linear) of the cementitious system, which occurs under sealed isothermal unrestrained conditions.

Bingham material - material which when subjected to shear stress behaves as an elastic solid until the yield stress is reached, after which there is a linear relationship between shear stress and rate of strain (flow velocity). The slope of this linear relationship is the plastic viscosity. Fresh cement paste and fresh concrete show such behaviour under certain conditions.

CDF test - standardized test method for the estimation of frost damage together with the use of a de-icing agent. CDF means capillary suction, de-icing solution, freezing [2].

Chemical shrinkage - internal, microscopical volume reduction, resulting from the fact that the absolute volume of the hydration products is smaller than that of the reacting constituents (cementing materials and water).

Chloride migration - ingress of chloride into the concrete pores/cracks from outer sources, such as a de-icing agent and/or sea water.

Coalescence - process in which two phase domains of the same composition combine together and form a larger phase domain.

Consistence - measure of ease by which fresh concrete can be placed into a form. It is the same as, and interchangeable with, consistency and workability.

Controlled-permeability formwork - formwork acting as a filter through which excess air and bleed water can escape, but the cement paste is retained in the body of the fresh concrete.

Copolymerization - process of synthesizing polymer molecules together in a chemical reaction to form three-dimensional networks of polymer chains.

Covalently cross-linked polymers - polymers whose chains are linked together by covalent bonds.

Curing – method used to maintain satisfactory moisture content and temperature after concrete is placed for a given period. The goal of curing is to improve concrete properties by promoting cement hydration and also minimizing shrinkage. Curing methods include water and sealed curing, external and internal curing.

Degree of hydration - ratio between mass of hydrated cement and mass of total cement in cement paste, mortar, or concrete.

Drying shrinkage - macroscopical dimensional reduction of hardened concrete volume due to evaporation of water or moisture loss to the outer environment.

External curing - most of the traditional curing methods applied from the surface of concrete (externally). The methods of external curing include water ponding, water spraying, wet burlap, plastic sheeting, curing compounds, and some others.

Filigree structure - structure built using a construction process that integrates factory-precast and field-construction technologies.

Gas-filled voids - empty spaces, usually filled with air, in a solid material such as concrete.

Geosynthetics - term that describes a range of polymeric products generally used to solve civil engineering problems.

Internal curing - incorporation of a component into the concrete mixture, which serves as curing agent. Internal curing can be classified into two categories: internal water curing (or water entrainment) and internal sealing.

Internal curing agent - material which stores water in concrete and releases it over time in order to support curing. SAP, for example, is an efficient internal curing agent.

Internal sealing - introduction into the concrete mixture of a curing agent, which is intended to delay or prevent loss of water present in the system.

Internal water curing (or water entrainment) - incorporation of a curing agent into fresh concrete serving as an internal reservoir of water, which can gradually release water as the concrete dries out. Internal water curing methods include the use of superabsorbent polymers (SAP), pre-wetted lightweight aggregates (LWA), normal-weight aggregates (NWA), wood-derived products, etc.

Methyl cellulose ethers - chemical compound derived from cellulose. It is a hydrophilic white powder in pure form and dissolves in cold water, forming a clear viscous solution or gel. It is typically used as a thickener and an emulsifier in various food and cosmetic products.

Nanoparticles - very small particle that behaves as a whole unit in terms of its transport and properties. Nanoparticles are sized between 1 and 100 nanometers.

Plastic viscosity - slope of the linear part of the relationship between shear stress and shear strain rate in a Bingham material.

Polycarboxilate superplasticizer - polycarboxilate-based polymer, with hydrophilic functional groups, specifically developed as an effective dispersant, fluidifier, and high-range water reducing agent for concrete. It is used for increasing the concrete workability without increasing water content, or for maintaining workability with a reduced amount of water.

Polyelectrolyte hydrogel - polymers whose repeating units bear an electrolyte group. These groups will dissociate in aqueous solutions, making the polymers charged.

Pores in concrete - sum of gel pores (diameter < 10 nm), capillary pores (0.01 to 100 μm), entrained-air pores (tenth of a millimeter), and natural air voids (order of mm). Pores created by SAP have the same size as entrained-air pores.

Restraint stresses - stresses due to imposed deformations or restraints on movement that can be caused by shrinkage (resulting in tensile stress), expansion (resulting in compressive stress) or temperature changes.

Rheology - study of flow of matter, primarily in the liquid state, but also as soft solids under conditions in which they respond with plastic flow rather than deforming elastically in response to an applied force.

Scaling - surface deterioration of concrete due to freezing and thawing. De-icing agents aggravate scaling considerably. Air-entrainment makes concrete much more resistant to surface scaling. SAP can act similarly to air-entraining agents.

Sealed curing - method of curing aimed to prevent exchange of moisture or any other substance between the cured material and the surrounding media.

Self-compacting concrete (am. Self-consolidating concrete) - fresh concrete that has the ability to flow under its own weight, fill the required space or formwork completely, and produce a dense and adequately homogeneous material without the need for compaction.

Self-desiccation - reduction in the internal relative humidity of a sealed system when vapor filled pores are generated. This occurs when chemical shrinkage takes place at the stage where the paste matrix has developed a self-supportive skeleton, and the chemical shrinkage is larger than the autogenous shrinkage.

Self-healing – mechanism by which the width of a crack in concrete diminishes with time. The phenomenon may have physical, chemical, and/or mechanical causes [3].

Shotcrete (or sprayed concrete) - concrete conveyed through a hose/nozzle system, which is pneumatically projected at high velocity onto a substrate surface. Wet shotcrete uses ready mixed concrete whereas dry shotcrete is a concrete without water to which water is added at the nozzle.

Slump-flow test - test method to determine the spread of self-compacting concrete using the standard slump cone without compaction. For an example of a testing regime, see reference [4].

Spacing factor - parameter related to the maximum distance of any point within the cement paste of a concrete from the periphery of an air void. It is used to estimate the fraction of paste within some distance of an air void.

Super absorbent polymer - cross-linked polyelectrolyte which starts to swell upon contact with water or aqueous solutions resulting in the formation of a hydrogel. These polymers are able to absorb up to 1500 g of water per gram of SAP. In engineering practice, SAPs are mostly based on cross-linked poly acrylic acid.

Thermoplastic elastomer - class of copolymers or a physical mix of polymers (usually a plastic and a rubber) consisting of materials with both thermoplastic and elastomeric properties.

Ultra-high-performance concrete - new generation of concrete made with fibre-reinforced cementitious materials with enhanced mechanical and aesthetic properties that far exceed those of common concrete used in construction. It has considerably high compressive strength that can exceed 250 MPa and flexural strength that can reach 50 MPa. It also has high durability, abrasion resistance, and chemical resistance.

V-funnel test - test method to determine the flow time of self-compacting concrete using a standardized V-funnel. For an example of a testing regime, see reference [5].

Water curing - method of supplying additional moisture to a material. It can be used also for preventing moisture loss.

Water-entraining agent - additive to freshly-mixed concrete used to entrain water into the cement paste system for the purpose of internal curing. Water may be entrained in different ways; for example, by use of SAP, forming water-filled voids in hardened concrete, or by use of pre-saturated porous lightweight aggregate.

Water migration - absorption and desorption of water by SAP in concrete. When dry SAP particles come into contact with water during mixing of concrete, they rapidly absorb water and form water-filled cavities. When the cement paste self-desiccates due to hydration, water is released from SAP by capillary pressure.

Water-regulating agent - additive to freshly-mixed concrete for controlling the amount and flow of water needed to obtain specific concrete properties, which may include workability, low shrinkage from internal curing, or high strength.

Water retaining agent - material with the ability to absorb and store water from the fresh concrete.

Water-to-cement ratio (w/c) - ratio between the mass of water and the mass of cement in paste, mortar, and concrete. When concrete contains SAP, one has to distinguish between $(w/c)_{tot}$ which is the ratio of total added water to cement, $(w/c)_e$ which is the extra water (or entrained water) stored in the SAP, and $(w/c)_{eff}$ which is the difference between the two. Also, $(w/c)_{eff}$ is considered as the water-to-cement ratio responsible for the creation of capillary pores. Finally, $(w/c)_e$ is also called $(w/c)_{IC}$ because it refers to the amount of water available to internal curing.

Yield point - point on the stress-strain or shear stress-rate of shear strain diagram, which corresponds to the change from elastic to a plastic behaviour of a solid material, or from a static to a dynamic behaviour of a Bingham fluid.

Yield stress - stress required to initiate plastic deformation or flow of a material, corresponding to the yield point.

2.2 Notation

BSE	back scattered electron
CDF	capillary suction, de-icing solution, freezing
CEM	cement
CLCCURS	closed-loop computer controlled uniaxial restrained shrinkage testing apparatus
C-S-H	calcium silicate hydrate
EDX	energy dispersive X-ray emission
GSE	geosynthetic engineering
HPC	high-performance concrete
HSC	high-strength concrete
IC	internal curing
ITZ	interfacial transition zone
LWA	lightweight aggregate
MIP	mercury intrusion porosimetry
NWA	normal-weight aggregate
OPC	ordinary Portland cement
SAP	superabsorbent polymer
SCC	self-compacting concrete or self-consolidating concrete
UHPC	ultra high-performance concrete

References

[1] EN 206-1 (2000) Concrete - Part 1: Specification, performance, production and conformity

[2] Setzer MJ, Fagerlund G, Janssen DJ (1996) CDF Test – Test method for the freeze-thaw resistance of concrete - tests with sodium chloride solution (CDF). Recommendation by RILEM TC 117-FDC. Mater Struct 29: 523–528

[3] RILEM Technical Committee 221-SHC Self-healing phenomena in cement-based materials, Draft of the State-of-the-Art Report

[4] EN12350-8 (2010) Testing fresh concrete – Part 8: Self-compacting concrete - Slump-flow test

[5] EN 12350-9 (2010) Testing fresh concrete - Part 9: Self-compacting concrete - V-funnel test

Chapter 3
Superabsorbent Polymers (SAP)

Stefan Friedrich

Abstract Superabsorbent polymers (SAPs) are able to absorb up to 1500 g of water per gram of SAP. Chemically speaking, SAPs are cross-linked polyelectrolytes which start to swell upon contact with water or aqueous solutions resulting in the formation of a hydrogel. This chapter gives an introduction into the production and specific properties of SAPs relevant for their use in concrete construction.

3.1 Introduction

Superabsorbent polymers (SAPs) are one of the most fascinating materials in modern polymer technology [1, 2]. These polymers are able to absorb up to 1500 g of water per gram of SAP (Figure 3.1). They were developed in the late 1980s. The first application of SAPs was in diapers. The market grew very quickly and reached ca. 1 Mio t/a in the last years. Today, the main market for SAPs is still the hygiene industry (baby diapers and adult care articles).

Chemically speaking, SAPs are cross-linked polyelectrolytes which start to swell upon contact with water or aqueous solutions resulting in the formation of a hydrogel. In the hygiene industry only SAPs based on cross-linked poly acrylic acid are used (Figure 3.2), which is partially neutralized with hydroxides of alkali metals, usually sodium.

Traditionally the market for SAPs is split into two parts: Hygiene industry and technical SAPs. The latter comprises all applications apart from hygiene products. Technical SAPs, which can be based on acrylamide and acrylic acid, are used for example in landscaping, cable isolation, fire fighting, food packaging. Common SAPs for the hygiene industry are hard white granulates with a particle size of app. 150 to 850 μm.

S. Friedrich (✉)
BASF Construction Chemicals GmbH
e-mail: stefan.friedrich@basf.com

V. Mechtcherine and H.-W. Reinhardt (eds.), *Application of Superabsorbent Polymers (SAP) in Concrete Construction*, RILEM State of the Art Reports 2,
DOI 10.1007/978-94-007-2733-5_3, © RILEM 2012

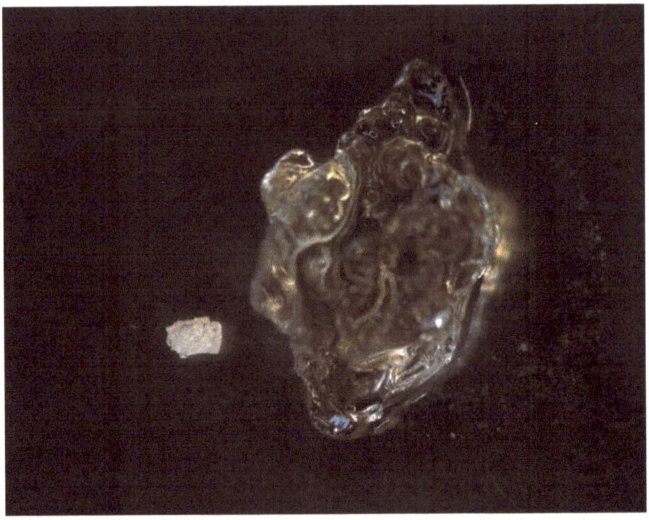

Fig. 3.1 Dry and swollen SAP particle (figure with courtesy of BASF)

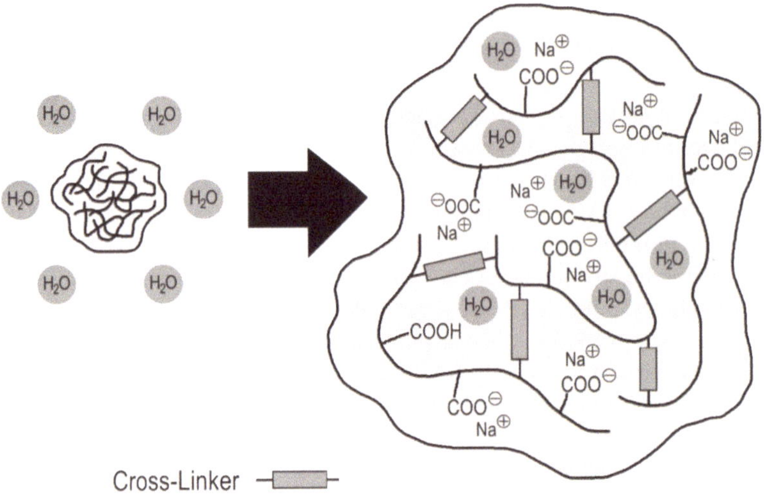

Fig. 3.2 SAP based on polyacrylic acid (figure with courtesy of BASF)

The main producers of SAPs for hygiene industries are BASF SE, Evonik Stockhausen GmbH and Nippon Shokubai. In the field of technical SAPs other producers include Arkema and SNF Floerger.

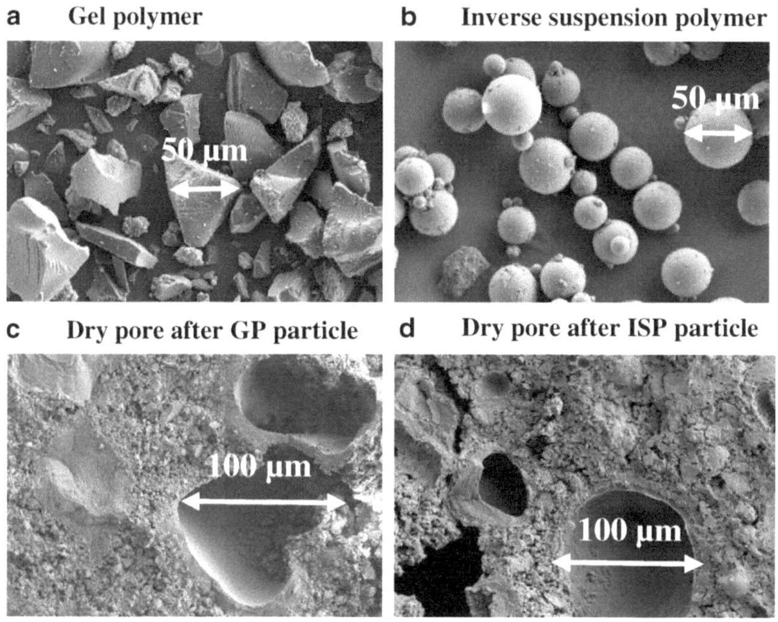

Fig. 3.3 Particle shape for polymers made by (**a**) gel polymerization or (**b**) inverse suspension polymerization; (**c**) dry pore left behind by a gel polymerized SAP particle in hardened cement paste after drying; (**d**) dry pore left behind in hardened cement paste by an SAP particle polymerized via inverse suspension polymerization (figures with courtesy of BASF)

3.2 Production

The production of SAPs starts with an aqueous monomer solution with a concentration of 25 to 40 mass per cent. The solution is cooled down to 0 to 10 °C and transferred to the reactor. This can be an endless belt reactor [3] or a kneader [4]. In the case of the endless belt, the monomer solution is poured out at the start of this belt and polymerization is running adiabatically forming a hard rubber-like gel. At the end of the belt an extruder cuts the gel into small pieces, which are then dried. The dry particles are ground to the desired particle size. In the case of the kneader the polymerization and the cutting of the gel are done in one step. Both processes are used on a large scale up to several 100.000 tons per year.

The particles which are made using these processes have an irregular shape and appear like broken glass under a microscope (Figure 3.3a).

An alternative production technology is inverse suspension polymerization [5, 6]. In this process the aqueous monomer solution is suspended in an organic solvent, e.g. hexane or cyclohexane. The polymerization is initiated between 50 and 70 °C and after the polymerization, water can be removed by azeotropic distillation. The product is filtered off and dried. SAPs which are made by this process are spherical (cf. Figure 3.3b). They can be single spherical particles or raspberry like agglomerates of smaller spherical particles.

The production of larger volumes of SAPs by inverse suspension polymerization is limited due to higher cost.

The swollen SAP particle keeps its particle shape. Figure 3.3 c) and d) shows the pores formed by SAP particles in cement paste after drying. The gel polymer forms irregular pores, the inverse suspension polymer spherical ones.

3.3 Swelling

The main driving force for the swelling of SAPs is the osmotic pressure which is proportional to the concentration of ions in the aqueous solution. As the ions in SAPs are forced closely together by the polymer network there is a very high osmotic pressure inside. By absorption of water the osmotic pressure is reduced by diluting the charges (Figure 3.2). The reset force of the polymer network and the external osmotic pressure are working to offset this osmotic driving force. Other external pressures, e.g. if the SAP has to swell or retain water against external mechanical forces, reduce the absorption capacity as well. When all forces are even the swelling is in equilibrium.

Therefore, the absorption of a SAP is strictly dependent on the concentration of ions in the swelling medium (Figure 3.4).

Di- and trivalent ions, e.g. Ca^{2+} and Al^{3+}, have an additional effect on the swelling behavior of SAPs which are based on polyacrylates. Because of their complex formation with carboxylate groups they act as additional cross-linkers dramatically reducing the absorption capacity (Figure 3.4).

It is possible to introduce other monomers containing ionic groups which do not form complexes with calcium or aluminum, e.g., sulfonic acids or cationic groups. In the latter case the network only contains cationic monomers [7]. However, such SAPs are not commercially available as yet.

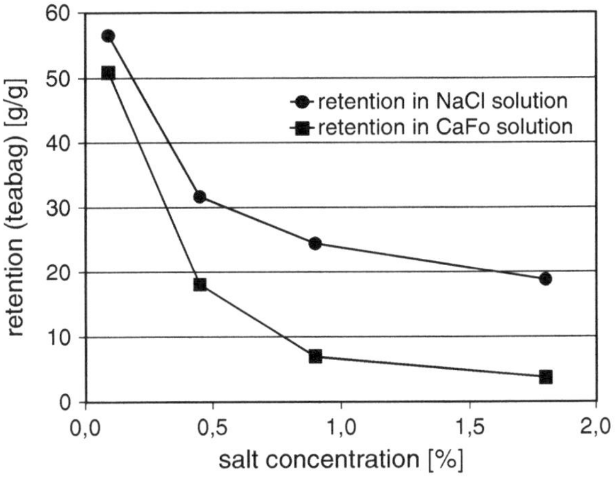

Fig. 3.4 Absorption capacity in sodium chloride (NaCl) and calcium formate (CaFo) containing solution (figure with courtesy of BASF)

Gel blocking is a special property of very fine SAPs having a particles size of less than 100 μm. If the SAPs are brought into contact water in pure form, little absorption takes place at the surface and the slightly swollen particles stick together. This results in lumps containing high amounts of not swollen SAP which do not disaggregate anymore. This effect forms the basis of the use of SAPs as sealing material. If fine SAPs are supposed to swell as individual particles, it is much more effective to distribute them before swelling, e.g. to blend the SAP with cement before mixing the mortar.

3.4 Characterization

Test methods for the characterization of SAP are standardized by EDANA, an association of the nonwoven industry [8].

A standard test for the absorption capacity is the so-called teabag method. A known amount of SAP is placed in a sealed teabag and put in a test solution. After the swelling time the teabag is removed, hung up to remove excess liquid and then weighed. The result is the absorption capacity. In the hygiene industry this is a very common method which has a high reproducibility. In a modification the teabag with the swollen SAP is put in a spin-dryer to apply a defined stress on the SAP.

In the hygiene industry it is common to use an isotonic NaCl solution, but in principle every test solution can be used, e.g. extracts of cement or artificial pore solutions.

A second widely used test is the absorption against external pressure. The test is called AAP (absorption against pressure) [9, 10] or AUL (absorption under load) [11]. The principle of this method is that the SAP has to swell and retain water against an external pressure.

Standard SAPs are modified to improve the AAP. After the grinding of the SAPs an additional cross-linker is sprayed onto the surface, in a so-called SX process. In this SX process a higher cross-linking density at the surface is created and a core-shell structure is formed. SX-products have a visual "drier" appearance on the surface in the swollen state.

For several technical applications the particle size distribution is very important. In former days it was determined using screen towers having several sieves sizes [12]. The result was a particle size distribution in fractions. For fine materials air-brush sieves were used. Today, more and more laser systems are used which can determine the complete Gauss distribution curve of a product [13].

Another parameter, which is strictly dependent on the particle size, is the speed of swelling. A quite simple test for the speed of swelling is the so called vortex-test. A defined amount of test solution is stirred in a beaker forming a vortex [14]. The SAP is added and the time until the vortex has disappeared is measured. This test can only be used for materials which do not show gel blocking.

For the determination of the absorption capacity or swelling parameters of fine particles it is possible to distribute the fine SAP in e.g. sand to avoid gel blocking. After the mixing time the SAP should not change the rheology by additional absorption. An alternative mechanism by which SAPs can affect the rheology over time is the release of extractables. These extractables are low molecular, not cross-linked parts in the SAPs which can migrate from the swollen SAP into the surrounding medium. These extractable parts can influence the rheology by thickening or they can retard the cement hydration. For the use in construction applications SAP having a low extractables (< 10%) should be used. For the determination of the extractable part extraction followed by titration of the polycarboxylic acid is used. If other monomers such as acrylamide are used, a determination of the total organic carbon in the extract is feasible.

As mentioned before the extractable parts are dependent on the particle size since the release in fine particles is easier because of the larger surface. However, extractables can also be formed during grinding processes as well as under mechanical stress. With high mechanical stress the polymer chains can be broken and extractables are formed.

3.5 Superabsorbents in Construction Applications

Right from the invention of SAPs the idea was to use them as additives for construction applications. The first patents were written by DOW and Hoechst dealing with dry mortars containing superabsorbents [15–17]. However, such products never were introduced into the market. At the end of the last century the focus was on internal curing of ultra-high performance concrete, which is described in the next chapters.

References

[1] Buchholz FL, Graham AT (eds.) (1998) Modern Superabsorbent Polymer Technology. Wiley-VCH, New York
[2] Frank M (2003) Superabsorbents. In: Bohnet M et al. (eds) Ullmann's Encyclopedia of Technical Chemistry, 7th edn, electronic release. Wiley-VCH Verlag
[3] Chmelir M, Pauen J (1988) Verfahren und Vorrichtung zum kontinuierlichen Herstellen von Polymerisaten und Copolymerisaten der Acrylsäure und/oder Methacrylsäure, DE 3544770 C2. Stockhausen GmbH, Krefeld, Germany
[4] Irie Y, Hatsuda T, Yonemura K, Kimura K (1996) Method of production of particulate hydrogel polymer and absorbent resin, EP 508810 B1. Nippon Shokubai Co., Ltd., Osaka, Japan
[5] Aoki S, Yamasaki H (1978) Process for preparation of spontaneously-crosslinked alkali metal acrylate polymers, US 4093776. Kao Soap Co, Ltd., Tokio, Japan
[6] Nakamura M, Yamamoto T, Tanaka H, Ozawa H, Shimada Y (1996) Process for production of water-absorbent resin, EP 441507 B1. Sumitomo Seika Chemicals Co, Ltd., Hyogo, Japan
[7] Schnee R, Masanek J, Fink H, Schleier W, Biedermann G et al. (1986) Schwach vernetzte, in Wasser schnell quellende, teilchenförmige feste Polymerisate oder Mischpolymerisate, Verfahren zu ihrer Herstellung und Verwendung in Hygieneartikeln, DE 3505920 A1. Röhm GmbH, Darmstadt, Germany

[8] http://www.edana.org. Accessed 23 March 2009

[9] Kellenberger SR (1993) Absorbent products containing hydrogels with ability to swell against pressure, EP 339461 B1. Kimberley-Clark Corporation, Neenah, Wisconsin, United States of America

[10] EDANA (2002) Recommended test method: Gravimetric determination of absorption under pressure, ERT 442.2 02

[11] Azad MM, Herfert N, Mitchel M, Robinson J (2003) Crosslinked polyamin coating on superabsorbent polymers, WO 2003/0436670 A1. BASF AG, Ludwigshafen, Germany

[12] EDANA (2002) Recommended test method: Particle size distribution – sieve fractionation, ERT 420.2-02

[13] EDANA (2002) Recommended test method: Determination of content of respirable particles, ERT 480.2-02

[14] Joy MC, Hsu W (2005) Superabsorbent polymer having increased rate of water absorption, WO 2005/063313 A1. Stockhausen Inc., Greensboro, North Carolina, United States of America

[15] Meyer WC (1989) A polymeric blend useful in thin-bed mortar compositions, EP 327351 A3, 1989 The Dow Chemical Company, Midland, Michigan, United States of America

[16] Girg F, Böhme-Kovac J (1996) Verdickermischungen für Baustoffe, EP 504870 B1. Hoechst AG, Frankfurt am Main, Germany

[17] Girg F, Böhme-Kovac J, Mann HM (1996) Zusatzmittelkombination zur Verbesserung der Verarbeitbarkeit von wasserhaltigen Baustoffgemischen, EP 530768 B1. Hoechst AG, Frankfurt am Main, Germany

Chapter 4
Kinetics of Water Migration in Cement-Based Systems Containing Superabsobent Polymers

Pietro Lura, Karen Friedemann, Frank Stallmach, Sven Mönnig, Mateusz Wyrzykowski, and Luis P. Esteves

Abstract Superabsorbent polymers (SAP) absorb pore solution during mixing of concrete and release it when cement paste self-desiccates or is exposed to drying. Knowledge of the kinetics of water migration in and out of the SAP is essential for understanding and optimizing internal curing of concrete. This chapter discusses absorption of pore solutions in SAP and desorption of the SAP, both in model systems and in cement paste or concrete. When experimental results about SAP are missing in the literature, results obtained with other internal curing agents are presented and the applicability to SAP is discussed. A final section is dedicated to modeling of internal curing of concrete by SAP and especially to modeling of water migration to and from the SAP.

P. Lura (✉)
Empa, Swiss Federal Laboratories for Materials Science and Technology, Dübendorf, Switzerland

Institute for Building Materials, ETH-Zurich, Zurich Switzerland
e-mail: pietro.lura@empa.ch

K. Friedemann • F. Stallmach
Universität Leipzig, Leipzig, Germany

S. Mönnig
BASF, Germany

M. Wyrzykowski
Empa, Swiss Federal Laboratories for Materials Science and Technology, Dübendorf, Switzerland

Technical University of Łódź, Poland

L.P. Esteves
Porto Engineering Institute, Portugal

V. Mechtcherine and H.-W. Reinhardt (eds.), *Application of Superabsorbent Polymers (SAP) in Concrete Construction*, RILEM State of the Art Reports 2,
DOI 10.1007/978-94-007-2733-5_4, © RILEM 2012

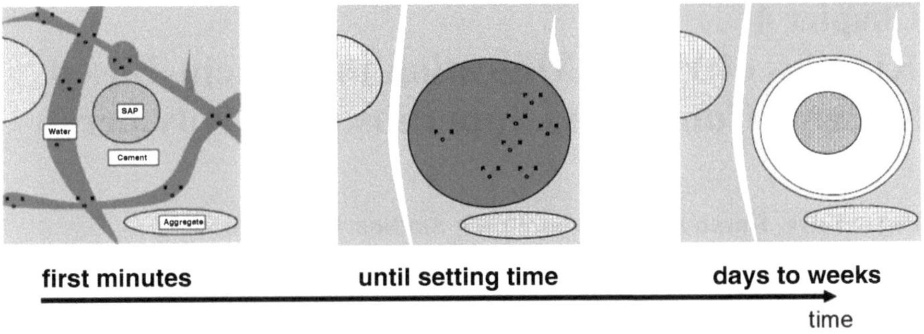

Fig. 4.1 Schematic representation of the evolution in time of the SAP in a cementitious material, after [2]. Left: initial condition, homogenous dispersion of cement particles, water, SAP and aggregates. Centre: the SAP has reached final absorption. Right: the water has been transported into the cementitious matrix and an almost empty pore remains

4.1 Introduction

Superabsorbent polymers (SAP) have recently found application in concrete technology, thanks to their ability to absorb amounts of water many times their own weight, retain it and release it when the conditions change [1]. In most applications, SAP have been added in the dry state to the concrete mixture. When dry SAP particles come into contact with water during mixing of concrete, they rapidly absorb it and form water-filled cavities (Fig. 4.1, left). The kinetics of absorption and the amount of fluid absorbed by the SAP depend both on the nature of the SAP and of the cement paste or concrete, in particular on the pore solution composition. Once the SAP have reached their final size, they form stable, water-filled inclusions (Fig. 4.1, center), from which the water is subsequently sucked into smaller capillary pores and consumed by hydration of cement. The SAP end up as empty pores in the cement paste (Fig. 4.1, right). This chapter discusses the process of absorption and desorption of the SAP in concrete. Section 4.2 is dedicated to absorption of pore fluid into the SAP, while section 4.3 deals with desorption. Finally, section 4.4 is dedicated to modeling of internal curing with SAP.

4.2 Absorption

4.2.1 Driving forces of absorption

Absorption of pore fluid into the SAP is the result of a competitive balance between expansive and shrinking forces. A high concentration of ions exists inside the SAP leading to a water flow into the SAP due to osmosis; another factor contributing to increase the swelling is water solvation of hydrophilic groups present along the polymer chain. On the contrary, elastic forces counteract swelling of the SAP [1]. More details on the SAP can be found in Chapter 3.

In addition to parameters depending on the SAP architecture, the ionic strength of the aqueous solution is of special importance for the swelling of the SAP. The ions in the solution change the inter- and intramolecular interactions of the polyelectrolytes due to shielding of charges on the polymer chain [1]. Especially the Ca^{2+} ions present in the pore solution of concrete can cause additional interlinking of the polymer chains and limit their swelling [2]. Furthermore, as the concentration of ions outside the SAP increases, the osmotic pressure inside the gel decreases, leading to a reduced swelling of the SAP [1].

The influence of particle size of the SAP should also be taken into account. According to Jensen and Hansen [3], very large SAP particles (a few hundreds μm across) may have a reduced efficiency due to insufficient time for water uptake during mixing. Very small SAP particles (a few μm across), on the other hand, may also show reduced absorption because of a less active surface zone compared to the bulk [3]. Recent work by Esteves [4] confirmed that the particle size of the SAP significantly influences both the amount of the pore solution absorbed and the rate of water absorption. Fick's second law of diffusion was used to describe the kinetics of absorption within a group of SAP depending on its particle size distribution [4].

4.2.2 Absorption in pore solutions

The concentration of different ions in the pore solution of cement pastes as a function of hydration time has been measured in several studies (e.g., [5]). The highest concentration, hundreds of mM, are normally found for K^+, Na^+, SO_4^{2-} and OH^- [5]. High concentrations develop immediately after mixing and remain roughly constant until setting time. Fig. 4.2 shows the ionic strength as a function of time for three cement pastes with water to cement ratio (w/c) 0.4, made with Portland cements differing in the alkali content [6]. At later ages, the ionic strength decreases as a consequence of precipitation; more important, the ionic strength differs considerably among Portland cements in dependence of their alkali content [6]. Addition of supplementary cementitious materials or use of alternative binders may influence the ionic strength even more.

Jensen and Hansen [3] measured absorption in a synthetic pore fluid ((mmol/l): $[Na^+]=400$, $[K^+]=400$, $[Ca^{2+}]=1$, $[SO_4^{2-}]=40$, $[OH^-]=722$) of two different types of SAP, both covalently crosslinked acrylamide/acrylic acid copolymers. A suspension-polymerized SAP (round particles, average 200 μm) absorbed about 20 g pore fluid/ g dry gel in 60 minutes, while a solution-polymerized SAP (crushed irregular particles, 125-250 μm), absorbed about 37 g pore fluid/ g dry gel in about 10 minutes [3]. Observations of SAP swelling in cement pastes (see section 4.2.3) indicate that the total absorption is about half the amount shown here for synthetic pore fluid [3].

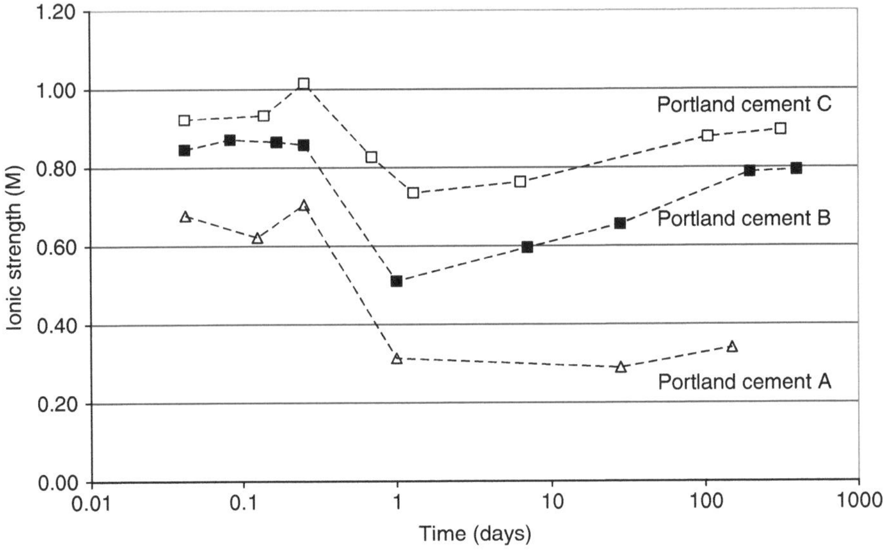

Fig. 4.2 Ionic strength of the pore solution of cement pastes (w/c 0.4) made with Portland cements differing in the alkali content, from low alkali (A) to high alkali (C) [6]. The ionic strength was calculated based on the measured pore solution composition as a function of hydration time

Pieper [7] measured the absorption of 11 different SAP in a simulated pore solution similar to the one used in [3]. Further 5 SAP types were tested in a synthetic pore solution based on the one extracted from Portland cement paste made at w/c 0.3 after 2 hours hydration at 20°C. Both the absorbent capacity for absorptive bound pore solution and the absorbent capacity for retentive bound pore solution (after applying vacuum for 3 minutes) was measured. According to [7], the ratio of the free absorbent capacity and the absorbent capacity under load provides a good prediction of the mechanical stability of the SAP particles.

Craeye and De Schutter [8] used SAP with particle sizes of 100 to 800 μm in the dry state. The water absorption after 24 hours was higher than 500 g/g, while after 300 seconds the water absorption was about 130 g/g. Based on this latter absorption level, the amount of SAP to be added to the concrete was estimated. It appears therefore that Craeye and De Schutter [8] may have not taken into account that the SAP may absorb much less in a pore solution than in water.

Esteves [4] observed with an optical microscope the growth in time of individual SAP particles in a synthetic pore solution. The particle size had a significant influence on both absorption kinetics and total amount of absorbed fluid. For instance, a particle with diameter of about 500 μm in the dry state absorbed almost 16 ml/g in 1 hour, while a particle with diameter of 50 μm micron absorbed 11 ml/g in less than 1 minute (Fig. 4.3).

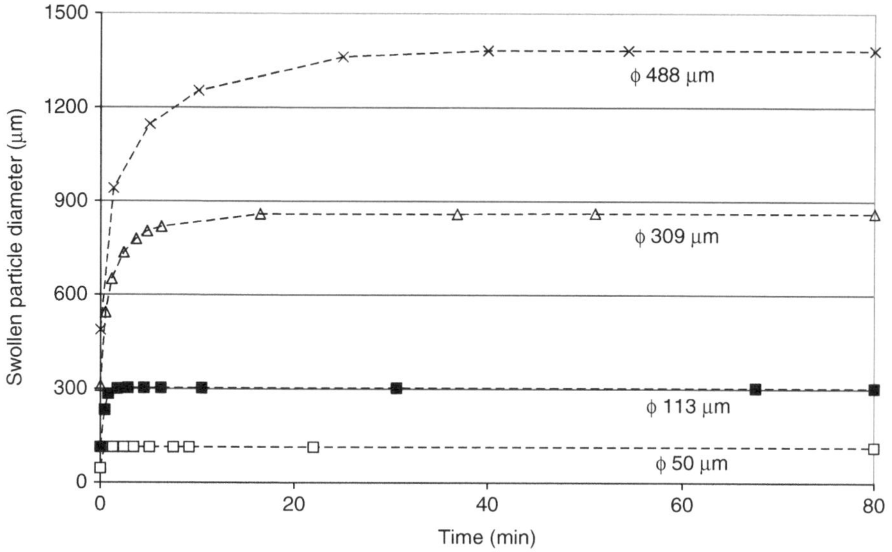

Fig. 4.3 Growth of individual polymer particles expressed by diameter change as function of time. The interaction medium is synthetic pore fluid. Adapted from [4]

4.2.3 Absorption in cement pastes

Compared to absorption in pore solution, determination of the absorption of pore solution by the SAP in a cement paste is difficult. Issues that have until now only received partial answers are the amount of fluid absorbed and the kinetics of absorption. No information is available about the differential swelling of different-sized particles in a cement paste; in particular, large particles may not have enough time to reach full expansion [3]. Additional points of interest are the distribution of the SAP in a concrete, in particular the formation of SAP clusters or agglomerations [2]. Finally, it should be investigated if vigorous mixing may crush the swollen SAP, especially the larger particles.

A number of authors estimated the absorption of the SAP by observing the changes in rheology of mixtures with SAP addition. Mönnig [2,9] estimated the absorption of two SAP types by comparing the slump flow as a function of additional mixing time (up to 15 minutes) of mixtures containing SAP. The SAP mixtures were compared with reference mixtures in which the w/c was varied stepwise. Absorption of 10 and 45 ml/g were found for two SAP types. Based on the change in slump flow with time, Mönnig [9] judged that most of the water uptake by the SAP occurred in the first 5 minutes. Dudziak and Mechtcherine [10] estimated the absorption of SAP in a coarse-grained fiber-reinforced Ultra-High Performance Concrete (UHPC) by adjusting the amount of water until they obtained the same slump flow of a reference mixture. Paiva et al. [11] investigated the rheology of single-coat renders with different SAP amounts, up to 0.15% of the total weight of the dry mortar. They observed an increase of both yield stress and plastic viscosity with the amount of SAP; however, no attempt was made to calculate the SAP absorption. More information on effect of SAP on fresh concrete properties is given in Chapter 5.

A method to estimate the absorption of an internal curing agent in a concrete mixture was developed by Johansen et al. [12] and applied to wood pulp fibers. A number of reference cement pastes with small w/c variations was compared to a mixture with an internal curing agent of unknown absorption. According to [12], the absorption was found when the rate of heat evolution of a reference paste coincided with the one of a paste with internal curing. However, it is difficult to reconcile this approach with evidence that a mixture with internal curing reaches a higher degree of hydration than a reference mixture [13] and is consequently also expected to develop a higher heat of hydration.

Based on the principles of stereology, the volumetric fraction of the pores produced by the SAP inclusions is equivalent to its area fraction determined by a random plane section. Therefore, to measure the amount of water absorbed into the SAP, it is sufficient to assess the area fraction of the voids in a random plane section. This approach was followed by Jensen [14], who prepared plane polished cross sections of water-entrained cement mortars with fluorescent paste and examined them by optical microscope. Both the amount of SAP and their distribution could be retrieved with this method. Similarly, Pieper [7] used a high resolution scanner to collect the digital images from a polished paste surface to quantify inclusion fraction. In addition, the particle size distribution of the dry SAP was obtained by image analysis of SEM pictures. Finally, absorption of about 7 g/g for a particular SAP type was calculated by comparing the particle size distributions in the dry and in the swollen state. Mechtcherine and co-workers [10,15,16] showed SEM pictures of SAP in the dry state and of spherical pores in the concrete resulting from SAP particles. They correlated the diameters of the dry SAP particles to the measured diameters sizes of the voids, finding that the SAP expanded about 3 times. Cryo-ESEM was recently employed to study morphology of hydrates in fresh cement pastes [17]. Cryo-ESEM may be used to examine cement pastes in the very first minutes during water uptake by the SAP, in order to study the kinetics of water absorption. It should be remarked that whereas the calculation of the total volume of the swollen SAP based on 2D images is straightforward, the same cannot be said for their particle size distribution. In fact, in the case of a spherical SAP, a random plane will on average not intersect the particle through its centre. Consequently, 2D images will show the SAP particles as circles with diameters that are on average smaller than the diameter of the corresponding spheres. Similar considerations can be made for irregular SAP.

Lura et al. [18] employed X-ray microtomography to study the particle size distribution of SAP and their three-dimensional distribution in high-performance mortars. The pore structure of mortars with SAP was characterized by round pores with a peak diameter at about 150 μm and was radically different from a plain mortar. A microtomography study of SAP in an ultra-high performance concrete was also performed in [10]; however, only an image of the 3D distribution of the voids was shown, without any quantitative analysis. A similar image was shown in [19], but no quantitative data were presented. More information on the pore size and distribution of SAP particles in concrete is given in Chapter 6.

Trtik et al. [20] used neutron tomography to observe the water uptake of a very large SAP particle (~1 mm in the dry state) inserted in a fresh cement paste with w/c 0.25. The absorption of the particle was about 12 g pore solution per g of SAP. The absorption process took about 3 hours, even if about 90% had taken place at 35 minutes after casting, when the first tomography was completed [20].

An alternative technique to study water absorption by SAP in fresh concrete is NMR relaxation. NMR experiments with high time resolution indicate a fast water uptake of SAP, within less than 5 min after water addition [21].

4.3 Desorption

4.3.1 Driving forces of desorption

When the cement paste self-desiccates due to hydration, a gradient in water activity is generated within the concrete between the water in the SAP and the pore fluid [22]. Part of this gradient of water activity is established by the capillary pressure developing in the pore fluid as a consequence of emptying of the pores due to hydration or external drying [23,24]. An additional contribution may be osmotic pressure, due to the fact that the composition of the pore solution in the cement paste at setting time and later may be different from that of the solution absorbed in the SAP. The process of desorption of the SAP may be described as a competition for water between the SAP and the cement paste [2].

4.3.2 Desorption of water and pore solutions

Jensen and Hansen [3] presented a desorption isotherm of a solution-polymerized SAP. While the free liquid uptake was 350 g/g in distilled water and 37 g/g in synthetic pore fluid (see section 4.2.2), at 98% RH the SAP retained only about 3 g/g of pore fluid and less than 1 g/g at 86% [3]. This may indicate that almost all the pore solution absorbed in the SAP will be released in the first days of hydration in a concrete with low w/c, where rapid self-desiccation takes place.

Mönnig [2] measured the desorption rate of SAP particles by monitoring the mass loss at different RH of layers of saturated SAP particles with a thickness of one particle; the SAP were saturated by spraying. This approach was necessary to avoid the complications arising with multiple SAP layers, where the results would

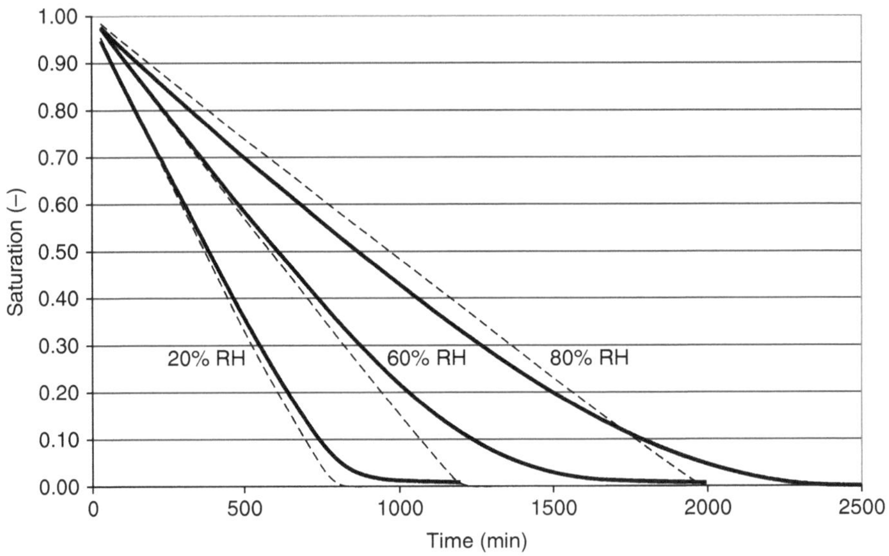

Fig. 4.4 Desorption measurements of SAP (thick, full lines) and plain water (thin, dashed lines) exposed to different RH levels at 20°C. The RH level is indicated next to the curves. Adapted from [2]

be influenced by the tortuous diffusion path between the particles. Results are shown in Fig. 4.4. After an initial period where desorption from SAP is similar to evaporation of free water, the curves depart from linearity. According to Mönnig [2], the water close to the surface of the polymers is lost rapidly but water closer to the core of the polymer must overcome more side-chains in the polymer, which interact with the water molecules through van-der-Waals forces.

4.3.3 Kinetics of desorption in cement pastes

The kinetics of desorption of the SAP in a cement paste depend on the properties of the SAP, on the kinetics of hydration, on the microstructure of the cement paste and on the interface between the SAP and the cement paste, through which the water transport takes place. Some experimental techniques described in the following section are able to quantify the amount of water that remains in the SAP (or is lost from them) at a given point during the hydration of a cement paste. However, a full understanding of the desorption process of the SAP, including the determination of the transport distance of water into the hardening cement paste, is still missing.

Mönnig [2] observed water transport from a single SAP in cement paste made at w/c=0.5 with an optical microscope. In this 2D experiment, a bright corona surrounding the SAP particle indicated a locally increased w/c. While the original cross section of the SAP was 168 μm, the influence zone had a diameter of 280 μm. This indicated a water transport distance from the SAP of at least 50-60 μm.

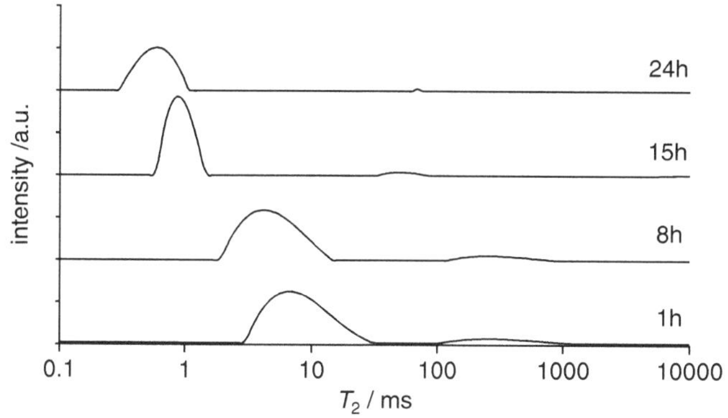

Fig. 4.5 NMR Relaxation time distributions of hydrating Portland cement paste of w/c 0.3 with internal curing by SAP (particle size 63-125 μm) calculated from inverse Laplace transformation of the measured transverse magnetization decay [21]. The main peak to the left is attributed to the physically bound water in the cement paste, while the smaller peak to the right represents the water inside the SAP. Between 8 and 15 h, a significant reduction in the content of water and in the relaxation time of the peak representing the water inside the SAP takes place, and after 24 h almost no water is detectable in the SAP [21]

Pieper [7] added a dye to water, which was then absorbed by SAP particles. During mixing of the cement paste, the SAP particles released part of the absorbed water. The aim of the investigation was to examine polished sections with light microscopy and to correlate the local intensity of the images with the transport distance of the internal curing water in the cement paste. However, the results were inconclusive.

Friedemann et al. [25] and Nestle et al. [21] used low-field NMR relaxation to study hydration in hardening cement pastes with and without SAP. Fig. 4.5 represents typical bimodal relaxation time distributions of an ordinary Portland cement paste (w/c 0.3) with addition of modified SAP after 1, 8, 15 and 24 h [21]. The significant peak observed at $T_2 \leq 10$ ms is attributed to the physically bound water in the hydrating cement paste and the small broad peak stretching from about 100 to 1000 ms in the freshly-mixed sample is assigned to the water absorbed in the SAP particles. Using a multi-exponential fitting routine, this SAP water is further split into two components. The corresponding two different water populations are assigned to surface relaxation at the interface between the cement paste and the SAP particles with incomplete diffusive exchange of water to the bulk gel [21]. The areas under the peaks measure the content of water in the respective environments and allow the calculation of the water uptake and of the swelling factor of the SAP. In the example of Fig. 4.5, the contribution of the curing water in the SAP to the total water content is about 15% [21].

By monitoring the time-dependent relaxation time distributions, the consumption of the physically bound water in the cement paste as well as the transition and consumption of the curing water are detectable. Figure 4.6 presents such data for a paste with SAP and for a reference paste. It is observed that the content of curing

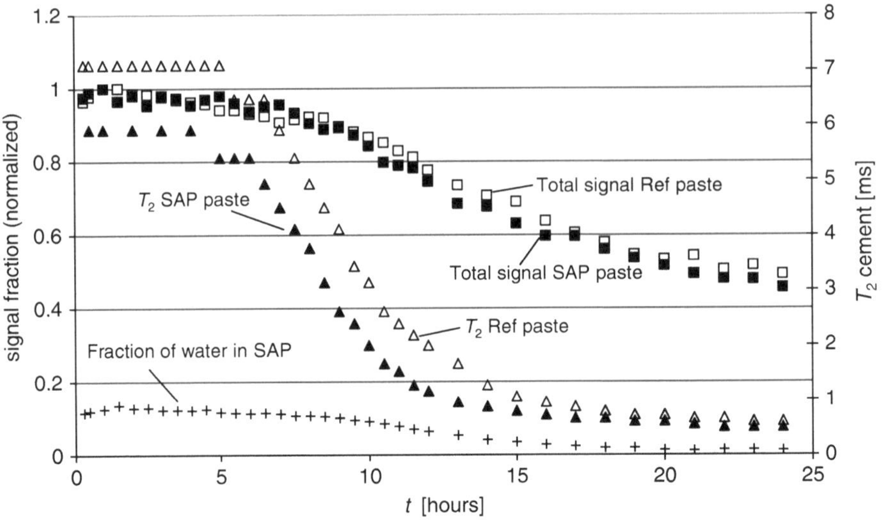

Fig. 4.6 Normalized signal intensities of the total signal, the water content in the SAP (particle size 63-125 μm) and the relaxation times T_2 of water in a Portland cement paste of w/c 0.3 for samples with internal curing by SAP (SAP paste) and for a reference sample (Ref paste). Data from Figure 11 in [21]

water inside the SAP decreases almost in parallel with the decrease of the water relaxation time in the cement matrix, while the decrease in the amplitude of the total water signal is still minimal. This may indicate that the water drainage from the SAP is mainly forced by the increase of capillary suction of the matrix which results from the increasing surface-to-volume (*S/V*) ratio. In fact, the relaxation rate $1/T_2$ is proportional to the *S/V*-ratio at this stage of the hydration. The water transfer from the SAP into the cement matrix is almost completed after 1 day. In Fig. 4.6 it is also evident that during the first hours of cement hydration, the relaxation times of the mixture containing SAP are shorter than that of the reference mixture. This may indicate the formation of a denser pore network with larger *S/V*-ratio in the presence of the SAP [21].

In addition, NMR relaxometry measurements allow the calculation of the degree of hydration from the time dependence of the NMR signal intensities [25]. In reference [25], the development of the water fractions of Portland cement pastes (w/c 0.275) containing SAP was analyzed and interpreted as function of the degree of hydration. By analyzing the content of water trapped in the SAP it was observed that the main transition of curing water (i.e., the desorption of the SAP) starts after the degree of hydration is about 0.1. At a degree of hydration of about 0.65, the water transition was almost completed in these cement pastes.

Pulsed field gradient NMR diffusion measurements were performed to study the water mobility of SAP particles as a function of water content, as well as of SAP inside hydrating cement pastes [21,26]. It was found that the water self-diffusion coefficients in SAP decrease slightly with increasing weight fraction of SAP, but remain high in the range of free water (10^{-9} m^2/s). This suggests that the diffusive

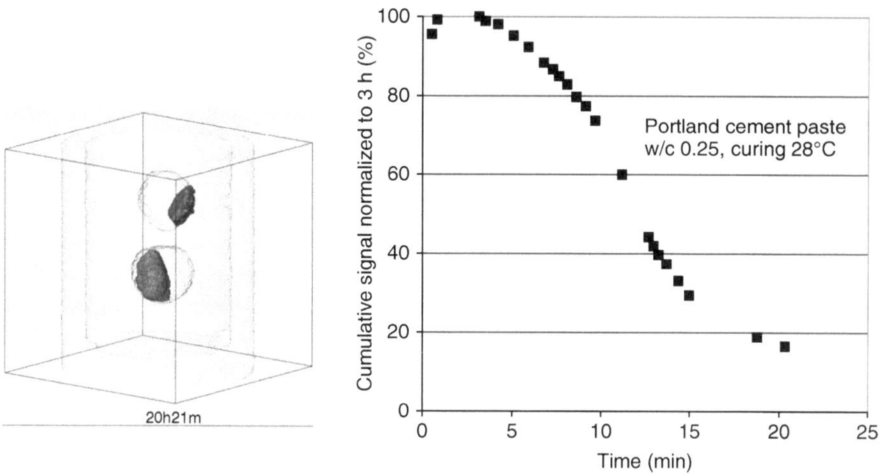

Fig. 4.7 Left: 3D representation of SAP particles within a hydrating w/c 0.25 cement paste sample at 20 hours after casting. The outer cylindrical shell is the Teflon mould, the inner cylinder is the cement paste and the two almost spherical particles are the SAP particles at the beginning of the experiment. The darker regions within the SAP particles represent the shrunk SAP particles at 20 hours after casting. Right: average signal from the SAP particles as a function of time, normalized to the signal at 3 hours. Adapted from [20]

exchange of water molecules between the SAP gel particles and the interface with the cement matrix is fast.

As described in [26], the diffusion length of (curing) water inside the hydrating cement can be estimated by the self-diffusion coefficients. In case of hydrating cement pastes of w/c=0.3, a self-diffusion coefficient of 5×10^{-10} m^2/s was measured after 10 hours. From this result, an average diffusion length of roughly 5 mm during 10 hours is calculated. This diffusion length determined by NMR is in good agreement with measurements of water transport from saturated lightweight aggregates (LWA) to cement paste [27,28,29].

With synchrotron X-ray microtomography it was possible to follow the emptying of LWA during internal curing of cement pastes [30] and its application to SAP appears straightforward. However, in [30] it was not possible to precisely quantify the distance of water transport into the cement paste.

Neutron radiography was recently employed to study water transport from saturated LWA to cement paste [31]. It may be applied to internal curing with SAP. The main disadvantage of neutron radiography is that a 2D geometry is required. Trtik et al. [20] applied neutron tomography to monitor the water release from large SAP particles (~1 mm in the dry state and ~2.5 mm in the swollen state) in a hydrating cement paste with w/c of 0.25 in the first day of hydration at 28°C. Very large SAP particles were employed in this study because of the resolution of neutron tomography, above 100 μm. The SAP started emptying around setting time and released about 80% of the water in the first day of hydration (Fig. 4.7). The data are currently being analyzed to determine the distance of water transport from the SAP.

From measurements of degree of hydration, heat of hydration, internal RH and autogenous deformation in cementitious systems with SAP, indirect evidence of the kinetics of water desorption from the SAP may be derived. In RH measurements [3] on cement pastes made at w/c=0.3, pastes contained 20% of silica fume and an amount of SAP required for optimal internal curing (0.6% by weight of cement, corresponding to an amount of entrained water of 0.075) did not show any self-desiccation in the first three weeks of hydration. On the contrary, pastes containing 0.3% of SAP showed a very moderate self-desiccation, but the internal RH remained substantially higher than in pastes without SAP. The development of autogenous deformation in the cement pastes followed closely the internal RH [3]. Similar results were obtained by Pierard et al. [32]. Lura et al. [13] measured higher non-evaporable water content at 28 days in cement pastes with SAP compared to reference pastes. Klemm [33] measured the rate of heat liberation on white Portland cement pastes with w/c 0.3 + 20% silica fume, either plain or with 0.6% SAP. The SAP did not appear to influence the dormant period of the paste, while both the height of the main hydration peak and of the secondary peak were increased by SAP addition. The SAP appeared to release water starting in correspondence of the main hydration peak and up to the first couple of days of hydration.

4.4 Modelling of Internal Curing with SAP

4.4.1 General on modelling of internal curing

Bentz et al. [34] reviewed applications of computational modeling to systems with internal water curing. Though not specifically targeted at internal curing with SAP, [34] discussed modelling approaches for issues relevant to this report, such as the distance to which water can effectively travel from the reservoirs and how the internal curing water is distributed within the concrete microstructure. This last issue was also addressed in [35], where a Hard-Core Soft-Shell Model [36] was used to calculate the amount of cement paste cured in a concrete with SAP.

4.4.2 Adapted DuCOM model

Mönnig [2] adapted the DuCOM model [37] to take into account the influence of SAP on degree of hydration and on a number of different concrete properties. The environmental conditions, the geometry of the specimen and the grid of elements were provided by a finite element code, while the degree of hydration, the saturation of the internal water sources, the pore size distribution and the water content of each element were calculated by a modified version of the DuCOM model. The fraction of cured cement paste was dependent on the amount of available curing water, on the distribution of the internal water sources and on the transport distance of the curing water. The internal water sources were homogeneously placed within the specimens, i.e., all elements had the same amount of internal curing water available at the beginning of the computation. The transport distances of the internal curing water were computed based on experimental data collected in [2]. The acting transport mechanism from the SAP to the paste depended on the RH (influenced by hydration and by drying of the paste) and on the saturation of the SAP. Two different transport mechanisms were accounted for: capillary suction and diffusion, with suction prevalent at high degree of saturation of the SAP and diffusion at low levels of saturation. The results showed that the surface/volume ratio of the SAP had the greatest influence on the water delivery rate to the concrete [2].

4.4.3 Two-scale modelling

A recently-developed mathematical model for hygro-thermal phenomena in maturing concrete by Gawin et al. [38,39] was applied to the two-scale modelling of internal curing of concrete.

A first approach to internal curing by SAP was described in [40]. The effective properties and macroscopic performance of the material were obtained by means of up-scaling the solutions from the mesoscopic level. At the mesoscopic level, the material was treated as a composite made of a single SAP inclusion surrounded by the maturing mortar [40]. At the mesoscale, the amount of water lost from the SAP was observed during hydration of the mortar and a function describing the water loss vs. hydration degree was determined. This function was then added to the macroscopic governing equation for water mass balance, acting as an additional mass term due to water source. The water and gas permeability and thermal conductivity of the concrete were determined based on the analysis of fluxes at the mesoscale and the other material parameters were volume-averaged. At the macro-scale, the real specimen was modelled by means of the finite element method, describing a strip cut-off of a cylindrical specimen [40], with the same dimensions and boundary conditions of the real experiment. Applications of the two-scale model are aimed in particular at simulating the macroscopic behaviour (e.g. self-desiccation and autogenous shrinkage, or drying shrinkage) of concrete during maturing. At the same time it is possible to observe the water desorption from SAP and migration kinetics in cement paste or concrete at the mesoscale [40]. As shown in a recent study [41], relatively accurate predictions of RH and autogenous shrinkage in mortars with SAP could be obtained with the model.

A further improvement of the two-scale model was described in [42,43], where additionally the different sizes of the SAP particles and their distribution in a cement paste were taken into account at the mesoscale (Fig. 4.8). This allowed for a more accurate analysis of water migration kinetics within the material and of the macroscopic transport properties of the porous medium.

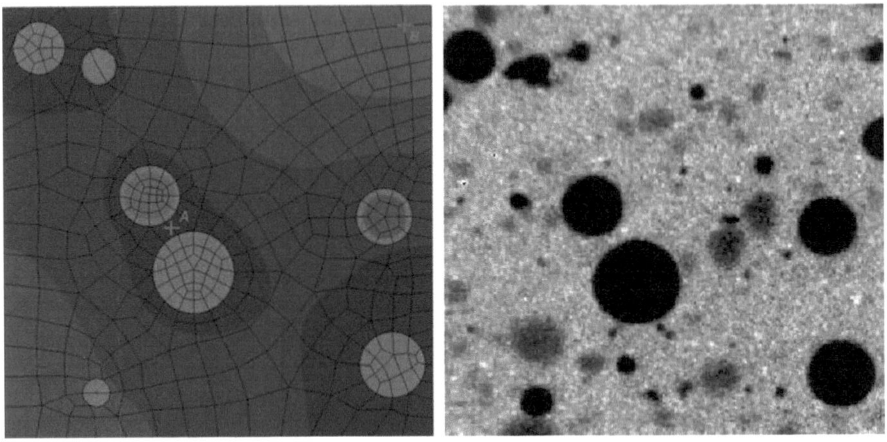

Fig. 4.8 Mesoscale results (left) concerning the saturation degree within the REV of a cement paste with SAP inclusions. The SAP are indicated as light gray disks; the decrease in saturation degree of the cement paste with growing distance from the SAP is indicated by lighter shades of gray. The distribution of the SAP inclusions was derived from a 2D section obtained from X-ray tomography (right). Adapted from [42]; X-ray tomography courtesy of Ye Guang

4.5 Conclusions

In this chapter, the current understanding about absorption and desorption of pore solution by SAP in cementitious materials was presented.

The behavior of SAP in water and in pore solutions was described and contrasted with the behavior in cement paste, mortar and concrete. The absorption and desorption in water and pore solutions appears to be well understood, while more data are needed about different types of SAP.

The kinetics of water transport in cement pastes have been addressed by a number of techniques. The common understanding is that the absorption of pore solution in the SAP is rapid and takes place in the first minutes after mixing. The water release from the SAP starts after setting time and appears to be completed after a couple of days, depending on the cement paste and on the curing temperature. In the opinion of the authors, NMR-based techniques, synchrotron X-ray tomography and neutron tomography seem to be very promising techniques to further clarify the mechanisms and the kinetics of water release in cement pastes and the distance to which internal curing water may travel. Data should be collected about the use of SAP with blended cements, possible interactions with other admixtures, and different mix compositions, especially in mortar and concrete.

Finally, a couple of recent approaches to numerical modelling of internal curing by SAP have been presented. It is expected that these models can be refined and other models developed as more information about the kinetics of water transport from the SAP becomes available.

Acknowledgements Thanks to Pavel Trtik (Empa) for critical reading of the manuscript.

References

[1] Jensen OM, Hansen PF (2001) Water-entrained cement-based materials – I. Principles and theoretical background. Cem Concr Res 31(4): 647–654

[2] Mönnig S (2009) Superabsorbing additions in concrete – applications, modelling and comparison of different internal water sources. PhD Thesis, University of Stuttgart, 164 pp

[3] Jensen OM, Hansen PF (2002) Water-entrained cement-based materials – II. Implementation and experimental results. Cem Concr Res 32(6): 973–978

[4] Esteves LP (2010) On the absorption kinetics of Superabsorbent Polymers. Int. RILEM Conf. on Use of Superabsorbent Polymers and Other New Additives in Concrete, 15-18 August 2010, Technical University of Denmark, Lyngby, Denmark

[5] Lothenbach B, Winnefeld F (2006). Thermodynamic modelling of the hydration of Portland cement. Cem Concr Res 36(2): 209–226

[6] Lura P, Lothenbach B (2010) Influence of pore solution chemistry on shrinkage of cement paste. The 50-year Teaching and Research Anniversary of Prof. Sun Wei on Advances in Civil Engineering Materials. C. Miao, G. Ye and H. Chen Eds., RILEM Publications SARL, 191–200

[7] Pieper M (2006) Innere Nachbehandlung von Beton mittels wasserspeichernden Polymeren (Internal Curing of Concrete via Superabsorbent Polymers). Aachen, Technische Hochschule, Fachbereich 3, Institut für Bauforschung, Diplomarbeit, (unpublished)

[8] Craeye B, De Schutter G (2006) Experimental Evaluation of Mitigation of Autogenous Shrinkage by Means of a Vertical Dilatometer for Concrete. Proc. Int. RILEM Conference - Volume Changes of Hardening Concrete: Testing and Mitigation, Eds. O.M. Jensen, P. Lura, and K. Kovler, RILEM Publications S.A.R.L., 21–30

[9] Mönnig S (2005) Water saturated super-absorbent polymers used in high strength concrete. Otto-Graf-Journal 16: 193–202

[10] Dudziak L, Mechtcherine V (2008) Mitigation of volume changes of Ultra-High Performance Concrete (UHPC) by using Super Absorbent Polymers. Proc. of 2nd Int. Symp. on Ultra High Performance Concrete, E. Fehling et al eds., Kassel University Press GmbH, 425–432

[11] Paiva H, Esteves LP, Cachim PB, Ferreira VM (2009) Rheology and hardened properties of single-coat render mortars with different types of water retaining agents. Constr Build Mat 23(2): 1141–1146

[12] Johansen NA, Millard MJ, Mezencevova A, Garas VY (2009) New method for determination of absorption capacity of internal curing agents. Cem Concr Res 39(1): 65–68

[13] Lura P, Durand F, Loukili A, Kovler K, Jensen OM (2006) Compressive strength of cement pastes and mortars with superabsorbent polymers. Int. RILEM Conf. Volume Changes of Hardening Concrete, 20-23 August, (Lyngby, Denmark), PRO 52, eds. Jensen, Lura and Kovler, 117–126

[14] Jensen OM (2005) Autogenous Phenomena in Cement-Based Materials, Department of Building Technology and Structural Engineering, Aalborg University, ISBN 87-91606-00-4, 188 pp

[15] Mechtcherine V, Dudziak L, Schulze J, Stähr H (2006) Internal curing by Super Absorbent Polymers – Effects on material properties of self-compacting fibre-reinforced high performance concrete. Int. RILEM Conf. Volume Changes of Hardening Concrete, 20-23 August, (Lyngby, Denmark), PRO 52, eds. Jensen, Lura and Kovler, 87–96

[16] Mechtcherine V, Dudziak L, Hempel S (2009) Mitigating early age shrinkage of Ultra-High Performance Concrete by using Super Absorbent Polymers (SAP). Creep, Shrinkage and Durability Mechanics of Concrete and Concrete Structures – CONCREEP-8, T. Tanabe et al. (eds.), Taylor & Francis Group, London, 847–853

[17] Zingg A, Holzer L, Kaech A, Winnefeld A, Pakusch J, Becker S, Gauckler L (2008) The microstructure of dispersed and non-dispersed fresh cement pastes — New insight by cryo-microscopy. Cem Concr Res 38(4): 522–529

[18] Lura P, Ye G, Cnudde V, Jacobs P (2008) Preliminary results about 3D distribution of superabsorbent polymers in mortars. Proc. Int. Conf. Microstructure-related durability of cementitious composites, RILEM Pro 61, 2008, Vol II, 1341–1348

[19] Jensen OM (2008) Use of superabsorbent polymers in construction materials. 1st Int. Conf. on Microstructure Related Durability of Cementitious Composites, Oct. 13-15, Nanjing, PRC
[20] Trtik P, Münch B, Weiss WJ, Herth G, Kaestner A, Lehmann E, Lura P (2010) Neutron tomography investigation of water release from superabsorbent polymers in cement paste. Int. Conf. on Material Science and 64th RILEM Annual Week Sept. 6-10, Aachen, Germany Volume III, PRO 77, 175–185
[21] Nestle N, Kühn A, Friedemann K, Horch C, Stallmach F, Herth G (2009) Water balance and pore structure development in cementious materials in internal curing with modified superabsorbent polymer studied by NMR. Micropor Mesopor Mat 125(1-2): 51–57
[22] Lura P, Jensen OM, Igarashi S-I (2007) Experimental observation of internal water curing of concrete. Mat Struct 40: 211–220
[23] Lura P, van Breugel K, Jensen OM (2003) Autogenous Shrinkage in High-Performance Cement Pastes: An Evaluation of Basic Mechanisms. Cem Concr Res 33(2): 223–232
[24] Weiss J, Lura P, Rajabipour F, Sant G (2008) Performance of Shrinkage-Reducing Admixtures at Different Humidities and at Early Ages. ACI Materi J 105(5): 478–486
[25] Friedemann K, Stallmach F, Kärger J (2009) Carboxylates and sulfates of polysaccharides for controlled internal water release during cement hydration. Cem Concr Comp 31: 244–249
[26] Friedemann K, Stallmach F, Kärger J (2006) NMR diffusion and relaxation studies during cement hydration – a non-destructive approach for clarification of the mechanism of internal post-curing of cementious materials. Cem Concr Res 36(5): 817–826
[27] Lura P, Bentz DP, Lange DA, Kovler K, Bentur A, van Breugel K (2006) Measurement of water transport from saturated pumice aggregates to hardening cement paste. Mater Struct 39: 861–868
[28] Henkensiefken R, Nantung T, Weiss J (2011) Saturated Lightweight Aggregate for Internal Curing in Low w/c Mixtures: Monitoring Water Movement Using X-ray Absorption. Strain 47(s1): e432–e441
[29] Trtik P, Münch B, Weiss WJ, Kaestner A, Jerjen I, Josic L, Lehmann E, Lura P (2011) Release of internal curing water from lightweight aggregates in cement paste investigated by neutron and X-ray tomography. Nucl Instrum Methods Phys Res A: Accelerators, Spectrometers, Detectors and Associated Equipment 651(1): 244–249
[30] Bentz DP, Halleck PM, Grader AS, Roberts JW (2006) Four-Dimensional X-ray Microtomography Study of Water Movement during Internal Curing. Int. RILEM Conf. Volume Changes of Hardening Concrete, 20-23 August, (Lyngby, Denmark), PRO 52, eds. Jensen, Lura and Kovler, 11–20
[31] Maruyama I, Kanematsu M, Noguchi T, Iikura H, Teramoto A, Hayano H (2009) Evaluation of Water Transfer from Saturated Lightweight Aggregate to Cement Paste Matrix by Neutron Radiography. Nucl Instrum Methods Phys Res A: Accelerators, Spectrometers, Detectors and Associated Equipment 605(1-2): 159–162
[32] Pierard J, Pollet V, Cauberg N (2006) Mitigating Autogenous Shrinkage in HPC by Internal Curing using Superabsorbent Polymers. Int. RILEM Conf. Volume Changes of Hardening Concrete, 20-23 August, (Lyngby, Denmark), PRO 52, eds. Jensen, Lura and Kovler, 97–106.
[33] Klemm S. Charakterisierung der Zementhydrataion mit einer in-situ Kombination von Wärmeflusskalorimetrie und chemischem Schwinden (in German). Internship Thesis, Empa, Switzerland and Technical University of Freiberg, Germany, 2009, 100 p.
[34] Bentz DP, Koenders EAB, Mönnig S, Reinhardt H-W, van Breugel K, Ye G (2007) Materials Science-Based Models in Support of Internal Water Curing. RILEM Report 41 Internal Curing of Concrete Eds. K. Kovler and O.M. Jensen, RILEM Publications S.A.R.L., 29–43
[35] Mönnig S, Lura P (2007) Superabsorbent polymers - An additive to increase the freeze-thaw resistance of high-strength concrete. Advances in Construction Materials, Symposium in Honor of Hans W. Reinhardt, 23-25 July (Stuttgart, Germany), Springer, Part V, 351–358

[36] Bentz DP (2010) Mixture Proportioning for Internal Curing with Lightweight Aggregates. http://ciks.cbt.nist.gov/~bentz/intcuring.html
[37] Maekawa K, Chaube R, Kishi T. Modelling of concrete performance. E&FN Spon, 1999
[38] Gawin D, Pesavento F, Schrefler BA (2006) Hygro-thermo-chemo-mechanical modelling of concrete at early ages and beyond. Part I: Hydration and hygro-thermal phenomena, Part II: Shrinkage and creep of concrete. Int J Num Meth Engng 67(3): 299–363
[39] Gawin D, Wyrzykowski M, Pesavento F (2008) Modeling Hygro-thermal Performance and Strains of Cementitious Building Materials Maturing in Variable Conditions. J of Building Physics 31(4): 301–318
[40] Wyrzykowski M, Gawin D, Pesavento F (2008) Determining the effective properties of internally cured concrete by means of two-scale modelling. Proc. 6th Int. Conf. Analytical Models and New Concepts in Concrete and Masonry Structures AMCM2008, Łódź, Poland, (on CD), 9–11 June
[41] Wyrzykowski M, Lura P, Pesavento F, Gawin D (2011) Modeling of internal curing in maturing concrete. Cem Concr Res 41: 1349–1356
[42] Wyrzykowski M (2010) Modeling coupled thermo-hygral processes in maturing concrete exposed to internal and external curing. PhD Thesis, Technical University of Lódź, Poland
[43] Wyrzykowski M, Lura P, Pesavento F, Gawin D (2011) Modeling of water release from superabsorbent polymers during internal curing. ASCE J Mat Civil Eng, accepted

Chapter 5
Effect of Superabsorbent Polymers on the Workability of Concrete and Mortar

Romildo D. Toledo Filho, Eugenia F. Silva, Anne N.M. Lopes, Viktor Mechtcherine, and Lukasz Dudziak

Abstract The use of Superabsorbent Polymers (SAP) to promote internal curing of concrete is one of the techniques used to mitigate its autogenous shrinkage. SAP absorbs water from the fresh mixture and releases it in a later stage when the relative humidity of the concrete pore system decreases due to the cement hydration. Several studies indicate that besides leading to a substantial reduction in autogenous shrinkage the addition of dry SAP to concrete also affect other properties of concrete such as workability, mechanical behavior and durability. This chapter gives an overview of the effect of SAP on the workability of High Performance Concrete (HPC), Ultra High Performance Concrete (UHPC) and OPC mortars. Considering that at present only few results are available in the literature regarding the influence of SAP on the workability of concrete, serious research efforts will be necessary to fully understand the rheological behaviour of concrete containing SAP.

5.1 Introduction

High performance concrete (HPC) and ultra-high performance concrete (UHPC) are materials characterized by exceptional mechanical properties and enhanced durability. However, they exhibit high autogenous shrinkage and therefore their early age cracking potential is high in applications where this deformation is restrained [1].

R.D.T. Filho (✉)
Civil Engineering Department, Federal University of Rio de Janeiro, Brazil
e-mail: toledo@coc.ufrj.br

E.F. Silva
Civil Engineering Department, University of Brasilia, Brazil

A.N.M. Lopes
FURNAS Hydropower Company, Brazil

V. Mechtcherine • L. Dudziak
Institute of Construction Materials, Technische Universität Dresden, Germany

V. Mechtcherine and H.-W. Reinhardt (eds.), *Application of Superabsorbent Polymers (SAP) in Concrete Construction*, RILEM State of the Art Reports 2,
DOI 10.1007/978-94-007-2733-5_5, © RILEM 2012

In order to reduce the autogenous shrinkage of concrete, several approaches have been studied in recent years. One of these possibilities is the use of internal curing that counteracts the adverse effects of internal dehydration by ensuring enough water for the continuation of the cementitious reactions [2]. Among the substances proposed as internal cure agents, SAP seems to be a promising one.

As explained in previous chapters, SAP consists of cross-linked chains which have dissociated ionic functional groups facilitating the absorption of large amounts of water. They are substances which can absorb many times their own weight of liquids by forming a gel. Due to this characteristic the rheological properties of fresh concrete containing SAP can change dramatically. Since the workability of fresh concrete expressed, for example, by its fluidity, compactability and ability to be pumped, depend on its rheological properties, a closer look is needed on the effects of SAP addition on the fresh concrete behavior.

The rheological behaviour of fresh concrete is most often described using equation (5.1) according to Bingham:

$$\tau = \tau_0 + \mu\dot{\gamma} \tag{5.1}$$

where τ shear stress applied to the material
$\dot{\gamma}$ shear strain rate
τ_0 yield stress
μ plastic viscosity.

In this model the yield stress and plastic viscosity are assumed to characterize the rheological behaviour of fresh concrete. These constants are normally obtained from tests carried out using concrete rheometers and mortar/cement paste viscosimeters. However, in many instances it has been observed that empirical tests are performed in order to obtain properties that can be associated with the rheological constants described above. For example, when testing Self-Compacting Concrete (SCC), the slump flow diameter, a parameter that describe the concrete flowability, is linked to the yield stress of fresh concrete whereas the flow-out time in the so-called V-funnel test is often correlated with the plastic viscosity of SCC.

The aim of the present chapter is to compile the results available on the literature about the workability of concrete and mortar containing SAP. The influence of some mixture parameters such as w/c, extra water for internal curing (due to the absorption of mixture water by the SAP) and the presence of admixtures on the concrete workability will be addressed as well.

5.2 Workability of Concrete and Mortar Containing SAP

The addition of SAP to concrete can change considerably its rheology. Jensen and Hansen [3] mention that the addition of 0.4% of a certain type of SAP relative to cement mass will lead to a lowering of the free w/c of 0.06. This change in the w/c will cause an increase in the yield stress and in the plastic viscosity of concrete. In

Fig. 5.1 Slump Flow test on fresh UHPC [5]

addition to this pure water binding effect, it is believed that a further increase in the yield stress and plastic viscosity will be caused by the physical presence of the swollen SAP particles [4]. At present, only few results are available in the literature regarding the influence of SAP on the workability of concrete. In the next sections a summary of these results are presented.

5.2.1 Workability of concrete using empirical test methods

A special UHPC was developed in 2006 for the construction of the FIFA World Cup pavilion in Germany. This structure, whose details are presented in Chapter 10, was designed as a filigree, thin-walled structure with very slender columns (minimum wall thickness of 20 mm) and without conventional reinforcement [5, 6].

Four UHPC mixtures met the design requirements regarding the strength and workability. Two mixtures (Ref and Pav) had no SAP addition and presented a mixture composition very similar with w/c of 0.24 and 0.25. The other two mixtures (Ref-SAP and Pav-SAP), also presenting mixture composition quite similar, contained 0.4% SAP related to the mass of cement. Extra water was added for internal curing to this mixes in the quantity sufficient to compensate the loss of the workability due to the absorption of mixture water by SAP. Consequently, the total water-to-cement ratios increased to 0.29 and 0.28, respectively. The mixture used for the construction of the pavilion was the Pav-SAP. Polycarboxilatether superplasticizer was used to attain a self-compacting behaviour of the concrete. A maximum aggregate size of 5 mm was chosen in order to comply with the geometry of the extremely filigree structure and the enhanced ductility of the UHPC was obtained using about 1.9% of steel fibers with a length of 6 mm and a diameter of 0.015 mm as disperse reinforcement [5].

The fresh properties of concrete were measured using the slump flow spread and the V-funnel time. The slump flow test was carried out using the standard slump cones for concrete (height of 300mm) and for mortars (height of 60mm). Figure 5.1 shows the details of the slump flow test using the 300mm height cone. Table 5.1 gives the results obtained from these tests.

Table 5.1 Results of rheological measurements [6]

Mixture	w/c total [-]	SAP [% m.c.]	Slump flow [mm] Small	Slump flow [mm] Large	V-funnel flow time [s]
Ref	0.24	0	240	675	44
Ref-SAP	0.29	0.4	265	765	14
Pav	0.25	0	270	770	5
Pav-SAP	0.28	0.4	270	780	13

Results show an increase in the slump flow and a decrease in the V-funnel flow time values of the mixture Ref-SAP when compared to those of the reference mixture (Ref). This behavior indicates that the amount of extra water used in the mixture Ref-SAP was not completely absorbed by the SAP. Mixtures Pav-SAP and PAV presented the same slump flow, even though Pav-SAP presents 0.03 of additional water. In this case, additional water for internal curing was completely absorbed by the SAP. A slight difference was observed, however, for the V-funnel flow time value that was higher for the mixture containing SAP.

The workability of fresh concrete containing SAP, measured by the slump flow test, was also determined in a study by Dudziak and Mechtcherine [7] in the framework of the Priority Program "Sustainable Building with Ultra High Performance Concrete". Two reference mixtures, a fine-grained UHPC enriched with steel fibers and a fiber-free UHPC containing coarse aggregates were tested. In addition, five UHPC mixtures containing SAP contents varying between 0.3-0.4% (related to the cement mass) were also studied. Additional water (0.04-0.07 related to the mass of cement) was used in the mixtures containing SAP.

For the fine-grained compositions, the slump-flow values ranged between 750 mm and 830 mm for the mixtures without and with the addition of SAP and extra water in their composition. Similar tendency was observed for the UHPCs containing coarser particles and fiber. The average slump flow value of the reference mixture was 680 mm whereas the UHPC with internal cure agent presented a slump flow of 695 mm.

In the continuation of their investigation, the authors extended the number of studied SAP-enriched mixtures as well as included additional mixture combinations [8]. Newly developed fine-grained UHPC mixtures were mixtures containing the amount of additional water equivalent to chosen mixtures with SAP. In turn, the workability of each SAP-enriched UHPC (F-S.4, F-S.3.04, F-S.4.07, F-S.6.08, F-S1.0.16 having 0.3-1.0% SAP and extra w_e/c=0.00-0.16 (see their comma-separated designations) was compared with that of the control mixture (nominated F-R) that had total water-to-cement ratio ($(w/c)_{tot}$) of 0.22 and no SAP or extra water in its composition. The workability of the mixtures F-S.4, F-S.3.04 and F-S.4.07 was also compared, respectively, to the workability of the mixtures with same water-to-cement ratio, i.e. F-R ($(w/c)_{tot}$=0.22), F-R.04 ($(w/c)_{tot}$=0.22+0.04) and F-R.07-2 ($(w/c)_{tot}$=0.22+0.07). The results obtained by the authors are presented in Figure 5.2a. Reference values of control mixture (0% SAP, extra w/c=0) produced with high intensity mixer and high intensity mixer equipped in vacuum unit are indicated

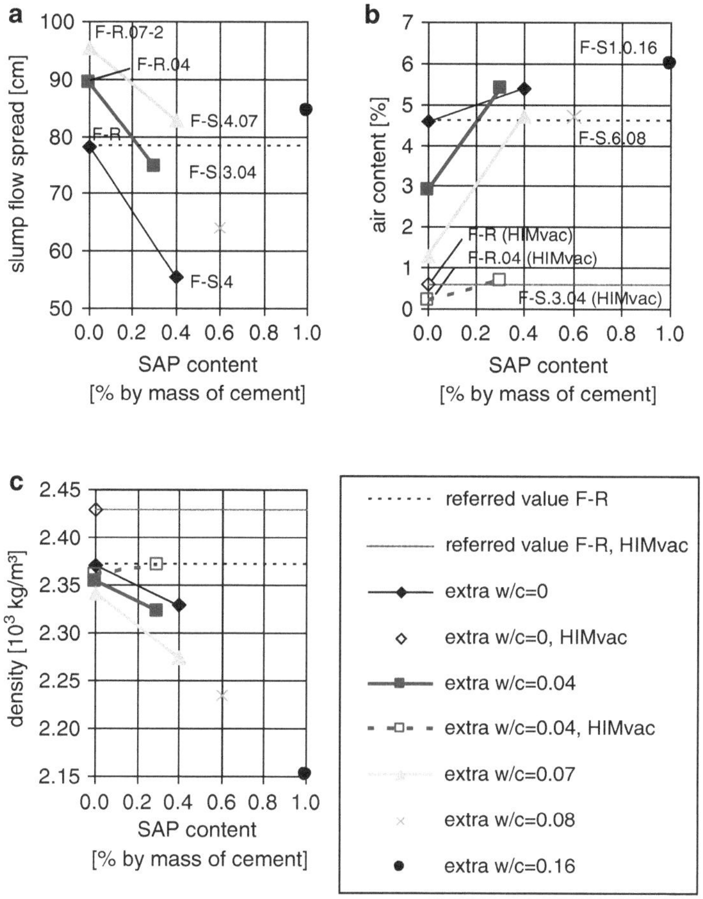

Fig. 5.2 Slump flow spread (**a**), air content (**b**), and density of fresh concrete (**c**) of UHPC mixtures depending on SAP content and extra w/c [8]

by the dashed black lines and grey dotted lines, respectively. In the graph, each line expresses the results obtained for identical overall w/c (including extra water) to better demonstrate the trends in the value changes due to the addition of SAP. The results indicate that in order to preserve the slump flow spread values presented by the control mixture (F-R) an extra w/c content of 0.015 has to be added for every 0.1% of SAP incorporated in the mixture. Such finding is drawn when one compares the slump flow spread values obtained from SAP-enriched mixtures with that of control mixture F-R and matched proportions of SAP and extra water used [8]. It is important to mention, however, that this conclusion is only valid for the particular type of SAP used in their study i.e. acrylamide/acrylic acid polymer, spherical shape of particles with diameter of approximately 150 μm in the dry state and water uptake of approximately 16±1 g as measured directly in concrete. When more water is added to the control mixture without the addition of SAP, its fluidity is dramatically increased. In fact, values typically recorded for homogenous self-compacting concrete are significantly exceeded. In this respect, there are no indications of loss of homogeneity by the mixtures containing SAP.

Regarding the air content of fresh concrete (see Figure 5.2b) the results indicate that all UHPC mixtures with SAP presented an increase in air content. This change was relatively low when comparison was made to the control mixture (i.e., the mixture without SAP and additional water, therefore having lower w/c), but pronounced when referenced to the mixture with additional water (equivalent to curing water in the corresponding SAP-enriched mixture). Regarding the density in the fresh state all UHPC mixtures with SAP presented some decrease in density (see Figure 5.2c). In fairly good agreement with these findings, study of Paiva et al. [9] showed that the addition of SAP and extra water resulted in a decrease in density and in an increase in the air content of fresh mortars.

In order to assess the absorption capacity of SAP under mixing condition, Mönnig [10, 11] studied the time evolution of the spreading flow of mortar mixtures containing varied contents of different SAP materials. The recorded slump flow measurements were expressed as a function of additional mixing time with regular 2 or 5 minute intervals. The corresponding results are illustrated in Figure 5.3a and Figure 5.3b.

It was pointed out by the author that progression of workability for SAP-containing mixtures resembles the one of reference mortar with particular amount of water. This finding held true when full saturation of SAP had been attained. Under this condition, precise information on water uptake was acquired. Interestingly enough, one may notice from the slope of curves in Figure 5.3a and Figure 5.3b between the first and second slump record that two relatively different SAP materials were tested (i.e., the absorption of polymer 1 was much longer). This suggests that some SAP materials may control the workability of cementitious materials for extended period of time (e.g., after the mixing process has been already finished). To what extent the implementation of SAP changes the rheological behaviour of mortar, cannot be derived from the studies carried out by the author. Not only slump flow is an insufficient tool for quantitative assessment of the rheological variables, but also the changes appearing between two corresponding mixtures without SAP and with curing agent after full saturation remained relatively small. In addition to this, the test duration was limited to 12-15 minutes after regular mixing whereas the SAP effects can be in fact observed for periods longer than the fluid state of cementitious material.

Mönnig [11] also reported the data concerning air content and density as recorded for numerous combinations of mortars and concrete. The study parameters were presence of SAP varied by type, particle size distribution and content as well as the magnitude of total w/c. Although SAP-enriched mixtures showed in general common tendency of increased air content and decreased density, signalized for previously discussed studies, some mixtures showed the opposite effect with regards to the former property. No conclusive explanation was given by author; however, it could be pointed out that some properties of polymers tested may potentially contribute to this result.

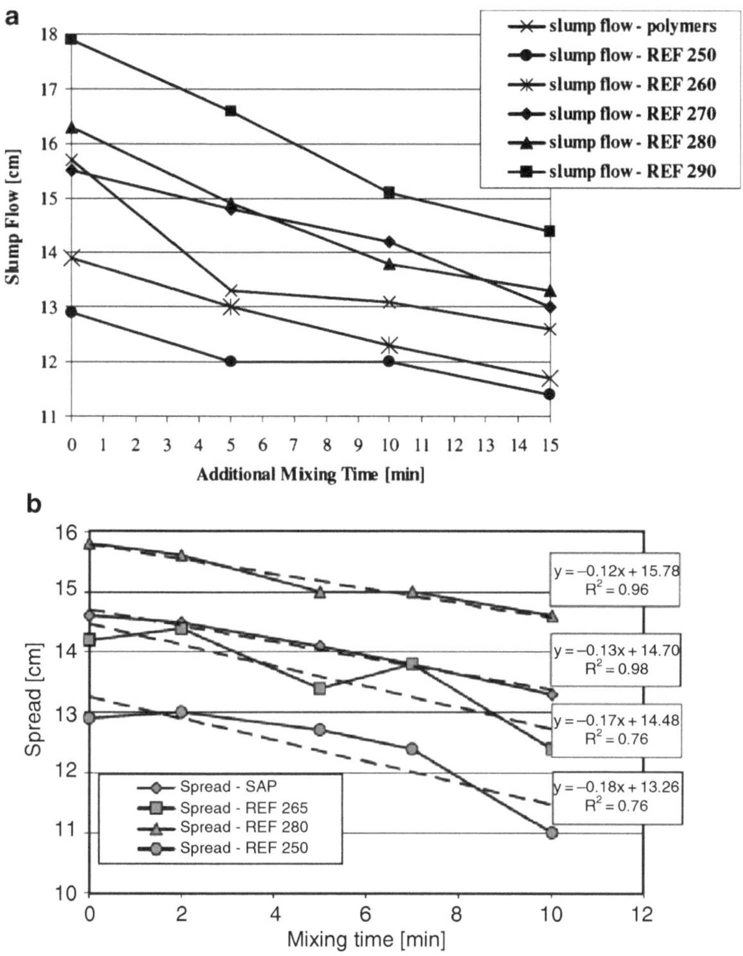

Fig. 5.3 Slump flow measurements of reference mortars, without any polymers, with different water content in comparison to mortars enriched with (**a**) polymer 1 [10] and (**b**) polymer 2 [11]

5.2.2 Rheological behaviour of concrete assessed from rheometer tests

Dudziak and Mechtcherine [12] examined the rheological behaviour of fiber-free finely grained UHPC by means of rheometer and V-funnel tests. The measuring procedure for the rheometer test included operating the equipment in alternate mode of rotation (6.5 min), oscillation (45 min) and again rotation (6.5 min). Flow time obtained from the V-funnel test was derived continuously until the age of approximately 1 hour at intervals of about 15 minutes.

Studied UHPC concretes were composed of the same basic ingredients as used in previous studies ([7, 8]). Designated for examination were control mixture F-R

Fig. 5.4 Torque variation for primary reference F-R and F-S.4.06 containing 0.4% SAP with angular frequency in first and second rotational mode [12]

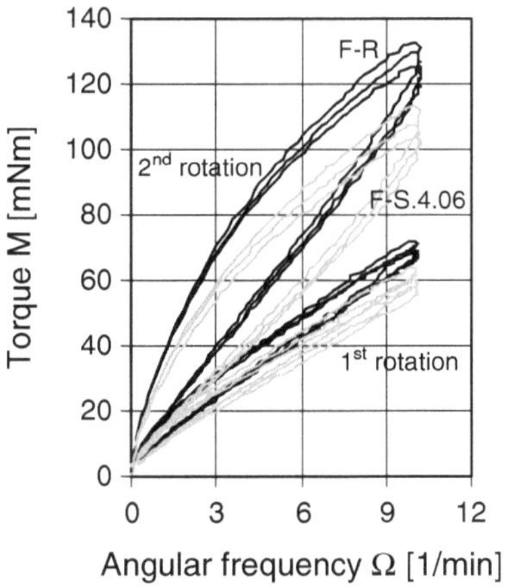

(0% SAP, w/c=0.22, slump flow measured with small cone 263 mm) and SAP-enriched mixtures containing extra water for internal curing: mixture F-S.3.045 (0.3% SAP, w/c=0.22+0.0345, slump flow 267 mm), mixture F-S.4.06 (0.4% SAP, w/c=0.22+0.06, slump flow 264 mm). Furthermore, mixtures containing 0.6% SAP and varied amount of extra water (as given in their designations) were tested: F-S.6.08 (slump flow 224 mm), F-S.6.085 (slump flow 251 mm) and F-S.6.09 (slump flow 277 mm).

Exemplary results for the first and second testing in rotational mode for mixtures F-R and F-S.4.06, fairly equal in measure of slump flow, are presented in Figure 5.4. It can be clearly recognized that the mixture with SAP addition was less viscous at both testing times. Results of the measurements in oscillation mode led to the same conclusion. Furthermore, the SAP-enriched mixture showed for all investigated time point after mixing lower values of V-funnel time in comparison to the reference mixture F-R, which was a finding consistent with the results from the rheometer tests. As a reason for this behaviour, authors indicated positive influence of spherical shape of each SAP particle that retains after absorption of water and produce a kind of ball-bearing effect. This effect might be further increased due to a higher air content in the mixtures with SAP. In conclusion, the effect of SAP addition was regarded as beneficial.

Figure 5.5 presents the results for mixtures with different content of extra water but the same amount of SAP. The presentation is limited here to the second rotational cycle, however in the first cycle a similar trend was observed. It was pointed out by authors that an underestimation of the amount of water absorption by SAP led to pronounced increase of torque, therefore pronounced loss of workability as already observed in previous studies. For instance, the loss of workability by mixture F-S.6.08 became so dramatic that the second rotational mode could not be processed. On the other hand, even a slight overestimation of the SAP absorption provoked a pronounced decrease in concrete viscosity, cf. mixture F-S.6.09.

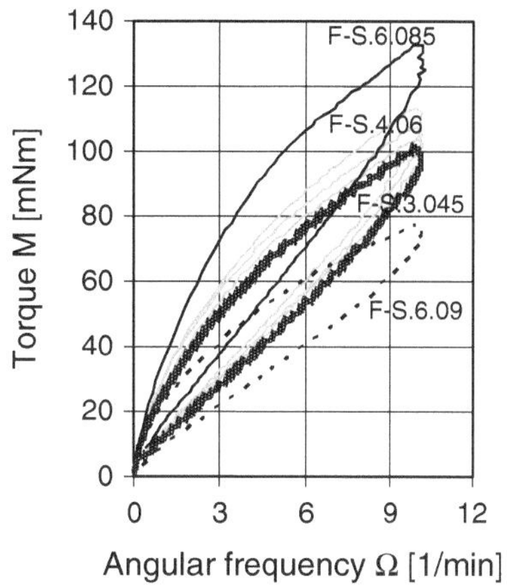

Fig. 5.5 Effect of SAP content on torque variation in second rotational mode [12]

5.2.3 Rheological behaviour of mortar assessed from rheometer tests

Paiva *et al.* [9] investigated the rheological behavior of single-coat render mortars containing SAP as a water retaining agent. A single-coat render is a mortar that is applied to an external wall surface in one layer. The hardened render must provide a durable weather resistant finish layer enhancing the performance of the building surface. The strength and durability depend on the absorbency by the support material and the environmental exposure at the moment the mortar is applied. It is important that the mortar retain the mixing water long enough to allow an adequate curing. The most commonly water retaining agents used are based in cellulose ethers. The rheological results obtained for the mortar containing commercial SAP are compared by the authors with those obtained using MHPC (a water retaining agent based in cellulose methyl–hydroxypropyl).

The single-coat render mortar basic composition involved white Portland cement and siliceous sand with a particle size distribution below 1.25 mm. Mortars were produced with a binder/aggregate weight ratio of 1:5 and water content of 21% by weight. The water retaining agents (SAP and MHPC) were added in powder form in contents of 0% (Ref), 0.05%, 0.08%, 0.10% and 0.15% of the total weight of the dry mortar. SAP had particle size distribution below 0.250 mm in the dry state.

The studied mortars should maintain their spreading in the range of 135–145 mm measured by the slump flow test using the mortar standard cone. The rheological properties (yield stress and a plastic viscosity) of the mortars were determined using a mortar rheometer (Viskomat PC, Schleibinger). The apparent density of the fresh mixtures was also determined.

Results showing the influence of SAP and MHPC on the values h (value related to the plastic viscosity and g (value related to yield stress) of the mortars are presented in Figure 5.6. SAP addition is equivalent to removing water from the system increasing

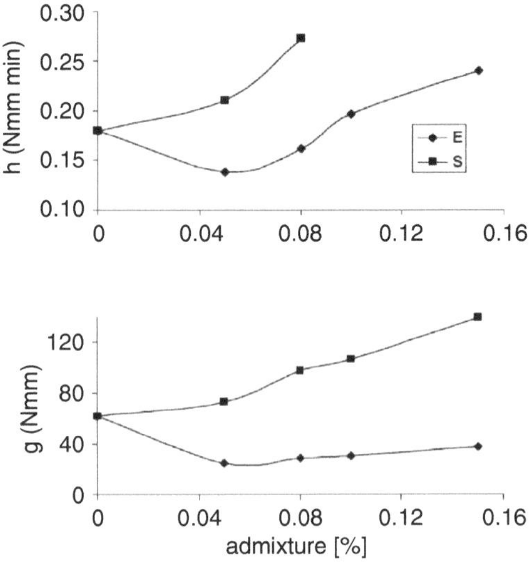

Fig. 5.6 Effect of admixtures content on the rheological parameters h and g of the mortar (0 minutes of test). Mixtures containing SAP are represented by “S” and containing MHPC by “E” [9]

the rheological parameter values and causing a thickening action. According to the authors [9], the increase in SAP content is equivalent to removing water from the system. For the same water retaining agent dosage, mixtures with MHPC presented smaller values of the plastic viscosity (cf. h) and the yield stress (cf. g).

5.3 Thickening effect caused by the SAP

According to Jensen [4], the thickening effect caused by the SAP can be used with advantage in some practical situations such as pumping. Particularly for shotcreting by the wet process, the thickening effect may be useful once a number of technological difficulties connected with this process can be avoided. To avoid these difficulties, the author tested in practice the potential use of dry addition SAP in the nozzle during shotcreting, promoting rheology modification. The uptake of water by the SAP created a viscosity change during placing and allowed the build-up of thick layers without the use of set-accelerating admixtures. More details about this practical application can be found in Chapter 10.

5.4 Final Remarks

The implementation of a new technique such as the use of SAP on concrete must require preliminary and extensive research in order to use the material with safety and effectiveness. At present only few results are available in the literature regarding the influence of SAP on the workability of concrete. In this chapter it was presented a summary of these results and it was seen that the addition of SAP to concrete can change considerably its rheology. It was observed, for example, that the addition of a certain type of SAP relative to cement mass is equivalent to removing water from the system. If extra water is not added to the mixture to compensate this tendency, an increase in the yield stress and in the plastic viscosity of the cementitious system is observed if a rheometer is used to measure the rheological properties. When empirical methods are used, a decrease in the slump flow spreading and an increase in flow time are observed. The presented results also indicate that mixtures with SAP presents an increase in its air content.

Further investigations will be necessary to fully understand the influence of SAP on the workability of concrete. It is believed that several topics should be addressed including the use of a concrete rheometer to determine the rheological properties of concrete mixtures containing SAP and large aggregates in its composition besides a better understanding of the kinetics of water absorption by SAP in fresh concrete and the subsequent water release in the hardening concrete.

References

[1] Mechtcherine V, Dudziak L, Hempel S (2009) Mitigating early age shrinkage of concrete by using Super Absorbent Polymers (SAP). In: Tanabe T et al. (eds) Proceedings of the 8th international conference on Creep, Shrinkage and Durability Mechanics of Concrete and Concrete Structures – CONCREEP-8, 30 September-2 October 2008 (Ise-shima, Japan), 847–853

[2] RILEM Report 41 (2007) State of the Art Report of RILEM Technical Committee Internal Curing of Concrete. Kovler K, Jensen OM (eds), RILEM Publications S.A.R.L., Bagneux, France, 141 pp

[3] Jensen OM, Hansen PF (2002) Water-Entrained Cement-Based Materials: II. Experimental observations. Cem Concr Res 32(6):973–978

[4] Jensen OM (2008) Use of superabsorbent polymers in construction materials. In: Sun W et al. (eds) Proceedings of the first international conference on Microstructure Related Durability of Cementitious Composites, 13-15 October (Nanjing, China), 757–764

[5] Mechtcherine V, Dudziak L, Schulze J, Stähr H (2006) Internal curing by Superabsorbent Polymers – Effects on material properties of self-compacting fibre-reinforced high performance concrete. In: Jensen OM, Lura P, Kovler K (eds) Proc of Int RILEM Conf on Volume Changes of Hardening Concrete: Testing and Mitigation, (Lyngby, Denmark), 87–96

[6] Dudziak L, Mechtcherine V (2008) Mitigation of volume changes of Ultra-High Performance Concrete (UHPC) by using Superabsorbent Polymers. In: Fehling E et al (eds.) Proceedings of the 2nd international symposium on Ultra High Performance Concrete, 5-7 March 2008 (Kassel, Germany), Kassel University Press GmbH, 425–432

[7] Dudziak L, Mechtcherine V (2010) Reducing the cracking potential of Ultra-High Performance Concrete by using Super Absorbent Polymers (SAP). In: Van Zijl GPAG, Boshoff WP (eds) Proceedings of the International Conference on Advanced Concrete Materials, 17-19 November 2009 (Stellenbosch University, South Africa), 11–19

[8] Dudziak L, Mechtcherine, V (2010) Enhancing early-age resistance to cracking in high-strength cement-based materials by means of internal curing using super absorbent polymers. In: W Brameshuber (ed) Proc Int Conf on Material Science (MatSci), 6-10 September 2010 (Aachen, Germany), RILEM Proc. PRO 77, Vol III, RILEM Publications S.A.R.L., Bagneux, France, 129–139

[9] Paiva H, Esteves LP, Cachim PB, Ferreira VM (2009) Rheology and hardened properties of single-coat render mortars with different types of water retaining agents. Constr Build Mater 23(2):1141–1146
[10] Mönnig S (2005) Water saturated super-absorbent polymers used in high strength concrete. Otto-Graf-J 16, 193–202
[11] Mönnig S (2009) Superabsorbing additions in concrete: applications, modelling and comparison of different internal water sources. PhD Thesis, University of Stuttgart, Germany
[12] Dudziak L, Mechtcherine V(2012) Effect of Superabsorbent Polymers' addition on the properties of Ultra-High Performance Concrete in fresh and hardened state (in preparation)

Chapter 6
Hardening Process of Binder Paste and Microstructure Development

Guang Ye, Klaas van Breugel, Pietro Lura, and Viktor Mechtcherine

Abstract Superabsorbent Polymers (SAP) added into high performance concrete prevents or reduces self-desiccation (or autogenous shrinkage) of concrete. The mechanism behind is free water release from saturated SAP leading to an increase of relative humility and promoting further hydration. Compared to plain cement paste, the addition of SAP changes the hydration process and the development of microstructure in concrete. In this chapter, the degree of hydration of cement in cement composites containing SAP is reviewed. The influence of SAP on the development of microstructure characteristics, i.e. porosity, pore size distribution, morphology and connectivity of bulk cement paste, interfacial transition zone between cement paste and SAP and the voids introduced by SAP is analyzed. It is concluded that the addition of SAP in the mixture increases the hydration degree of cement particle, this leads to a reduction of capillary porosity in the matrix. The additional of SAP meanwhile increase the void in the ITZ between SAP and matrix. Micro-computer tomography image shows that the voids introduced by SAP in the mixture are homogenously distributed.

6.1 Introduction

The main purpose of Superabsorbent Polymers (SAP) added into concrete is to prevent or reduce self-desiccation (or autogenous shrinkage) of concrete [1,2]. The mechanism behind this reduction is free water release from saturated SAP leading

---

G. Ye (✉) • K. van Breugel
Microlab, Delft University of Technology, the Netherlands
e-mail: G.Ye@tudelft.nl

P. Lura
EMPA, Dübendorf, Switzerland

V. Mechtcherine
Institute of Construction Materials, Technische Universität Dresden, Germany

V. Mechtcherine and H.-W. Reinhardt (eds.), *Application of Superabsorbent Polymers (SAP) in Concrete Construction*, RILEM State of the Art Reports 2,
DOI 10.1007/978-94-007-2733-5_6, © RILEM 2012

to an increase of relative humility and promoting further hydration. Compared to plain cement paste, the added SAP changes the kinetic of the hydration process and the development of microstructure in concrete. In this chapter, the degree of hydration of cement in cement composites containing SAP is reviewed. The influence of SAP on the development of microstructure characteristics, i.e. porosity, pore size distribution, morphology and connectivity of bulk cement paste, interfacial transition zone between cement paste and SAP and the voids introduced by SAP is summarized and discussed.

The experimental techniques used for testing the hydration process and the development of microstructure includes non-evaporable water content measurement, analysis of SEM-images and mercury intrusion porosimetery. The morphology and the connectivity of capillary porosity and voids were examined by means of computer microtomography.

6.2 Degree of Hydration of Cement Paste

The water released from SAP has a function to continue the hydration of cement further in sealed condition. The rate of the change of the degree of hydration depends on several material parameters, such as the size and the amount of SAP in the mixtures, the original water-to-cement ratio and the procedure how SAP was added in the concrete or mortar mixture [3]. The mechanisms of the water uptake of SAP in the cement paste and the rate of water release from SAP directly control the further hydration of cement (see Chapter 4).

Up to now, only a few studies were reported about the influence of SAP on the hydration process [4–6]. From backscattering electron image analysis, Igarashi and Watanabe [5] also observed no difference on the degree of hydration for cement paste made with w/c = 0.25 till the age of 14 days (Fig. 6.1). However,

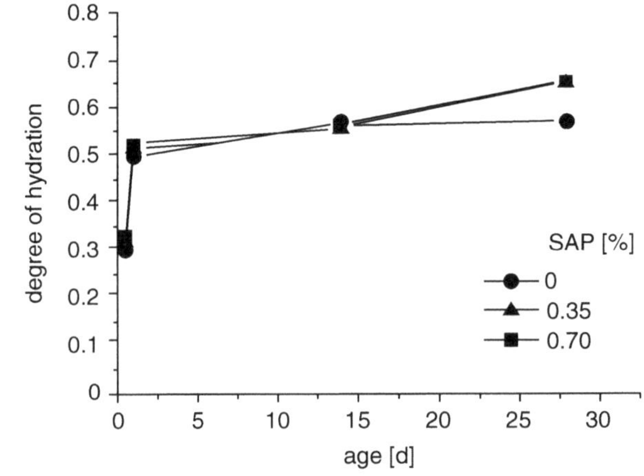

Fig. 6.1 Degree of hydration in cement pastes with SAP (w/c = 0.25) [5]

after 14 days the hydration degree in the cement pastes with SAP increased. Different content of SAP added, i.e. 0.35% and 0.70%, showed no difference on the hydration degree (Fig. 6.1), while no extra water was added in the mixtures. At age of 56 days, the cement paste with SAP the increase the relative degree of hydration by 10% was reported by Lura et al., which measured non-evaporable water content [6]. The same tendency was also found in the cement paste mixture with silica fume at 28 days.

Esteves [7] measured the degree of hydration of cement pastes and mortars containing silica fume by TGA-DTA, with and without water entrainment by means of SAP. The SAP promoted both the cement hydration and the pozzolanic reaction, especially in the first few days of hydration.

6.3 Pore Structure

Compared to normal concrete or high performance concrete, the presence of SAP certainly changes the microstructure in concrete. This section provides the information about corresponding findings. The main reasons causing the change of the microstructure, especially the pore structures in concrete containing SAP, are likely to be:

i. When SAP is fully filled with water, it acts as soft aggregate. When it is empty, SAP acts as air void in the concrete.
ii. Non-uniformity of the dispersion of SAP during mixing.
iii. Water uptake of SAP changes the effective w/c in the early hydration stage and water release from SAP affects the further hydration of cement.
iv. Interface between SAP and cement paste matrix may induce some additional pores.

The effect of SAP on the pore structure is not only on the total porosity, but also on the pore size and pore size distribution.

6.3.1 Total porosity

SAP increases the total porosity of concrete due to the hollow voids introduced by SAP particles, as it is shown by Mechtcherine et al. [8]. When 0.6% of SAP by mass of cement is added in the concrete mixtures, the total porosity measured by MIP at all ages is about 6% higher than that in plain concrete and 2% higher than that in the mixture with 0.3% SAP (Fig. 6.2) [8]. Note that extra water was used in this study in SAP-enriched mixtures in order to compensate the loss of workability due to the water absorption by SAP particles. Therefore, the total water-to-cement ratio was different in these mixtures. The MIP results reported by Mönnig [9] show no

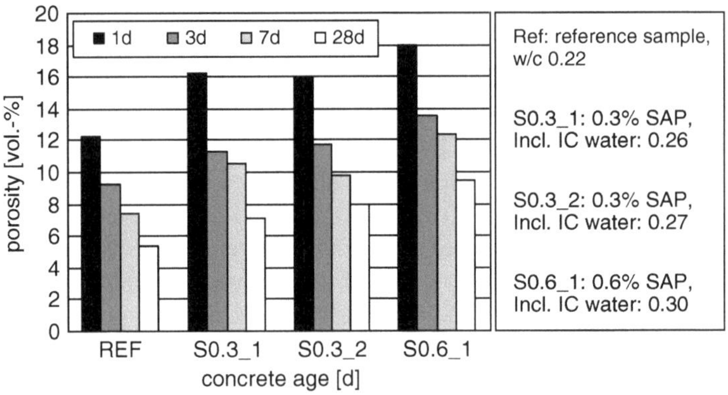

Fig. 6.2 Development of porosity as a function of UHPC composition and age [8]

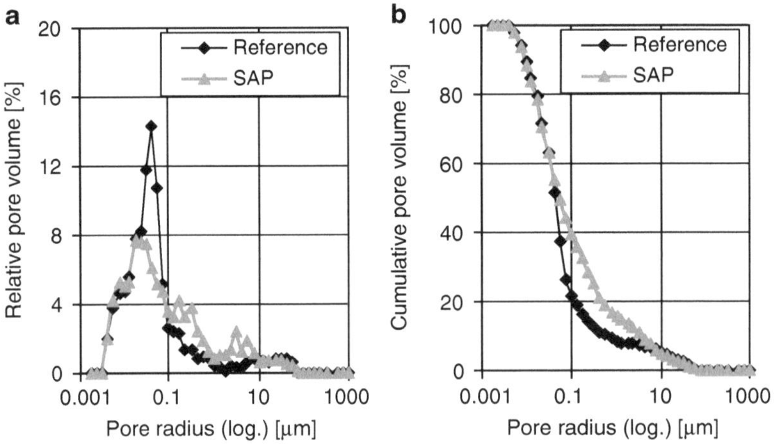

Fig. 6.3 Pore size distribution of concrete made with SAP addition and reference sample (w/c = 0.45, age = 1 day). After [9]

difference or even lower total porosity in the SAP mixture comparing to the reference mixture at 1 day (Fig. 6.3). However, in the investigation by Mönnig no extra water was used to compensate the change in rheological properties.

It has to be noticed that the pore diameters measured by MIP range from a few nanometer to about 120 μm. Since the average particle size of SAP is about 150 μm [8], the hollow voids introduced by SAP in concrete should be in average bigger than 150 μm (see also Fig. 6.10). Therefore, MIP technique could reveal only a part of these voids (those with smaller size).

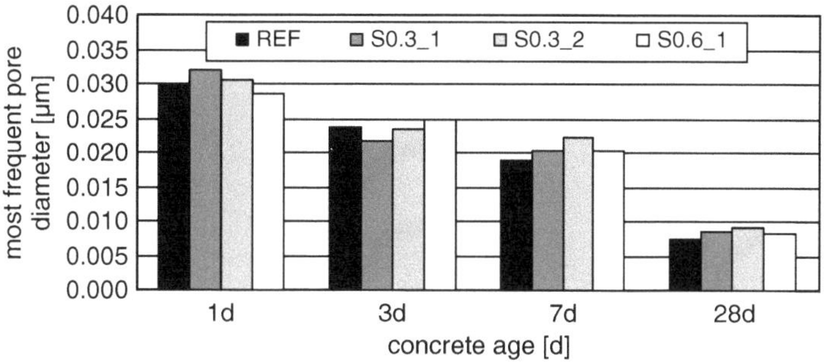

Fig. 6.4 Pore size distribution of UHPC made with and without SAP (in figure, S0.3, S0.6 indicate the amount of SAP by mass of cement in the mixture) [8]

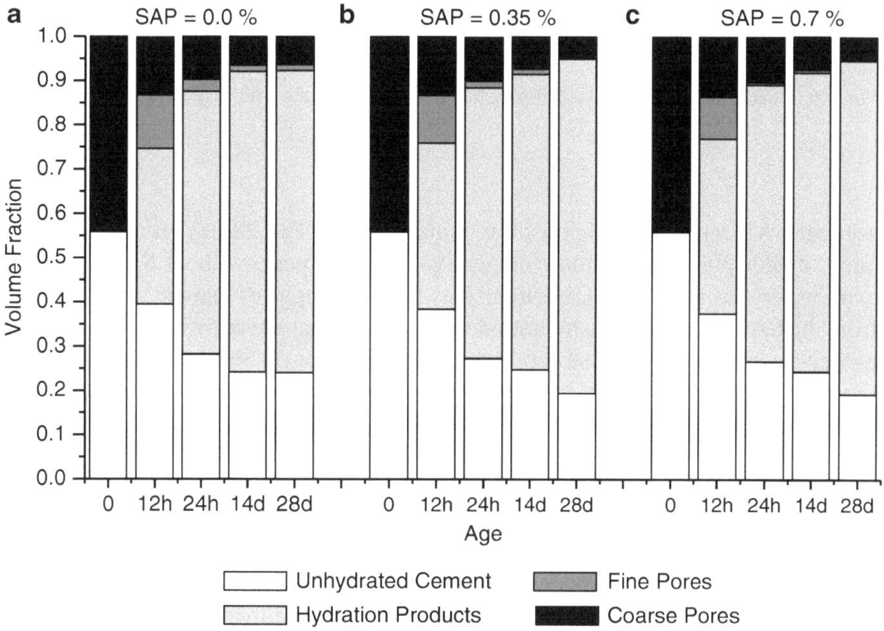

Fig. 6.5 Volume fractions of constituent phases in cement paste matrix surrounding SAP (w/c = 0.25) [5]

6.3.2 Pore size and pore size distribution

The SAP addition can change the pore size and pore size distribution of the surrounding cement matrix in a certain way. However, using MIP measurements Mechtcherine et al [8] could not find any significant effect of the SAP addition on the most frequent pore diameter in concrete (cf. Fig. 6.4). This holds true for the investigated ultra-high performance concrete at different ages.

Figure 6.5 shows the results from a back scatter electron (BSE) image analysis on volume fractions of constituent phases in the cement pastes made with and

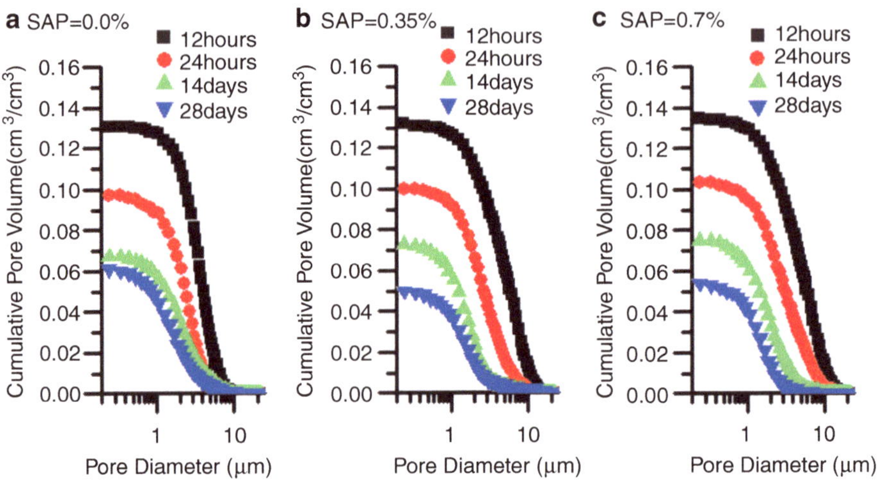

Fig. 6.6 Coarse capillary pore size distributions in cement pastes with SAP (w/c = 0.25) [5]

without SAP reported by Igarashi & Watanabe [5]. The 28-day pastes with SAP have smaller porosity and more cement gel than the paste without SAP. Note that extra water was used in this research. This fact also suggests that the water released from the SAP continued the hydration of cement so that the porosity was decreased, especially in the range of fine capillary pores at later ages. Similar results were also found by Dudziak et al [14] in ultra high performance concrete samples with SAP, where extra water were used in the mixtures.

Figure 6.6 shows coarse capillary pore size distribution curves obtained by BSE image analysis [5]. The diameter in the pastes without SAP changed little after 24hours while the diameter in the SAP-containing pastes appreciably decreased with time. Igarashi & Watanabe [5] suggested that the addition of saturated SAP particles decreased the volumes of large pores in the coarse pore structures.

In Fig. 6.7, it may be seen the combined effect on the hydration degree of silica fume added and water-entrained added systems, which is brought out by the direct comparison of the non-hydrated phase in 2D sections. This is supported by TG/DAT analysis to the same systems [7].

6.4 Interfacial Zone

As observed by Lam [4] from scanning electron microscopy (SEM) by using a BSE detector, mortar that contained the SAP had empty voids that were the remains of the spherical SAP particles (Fig. 6.8). These voids appear as black spots with a dark grayish rim. Some of the cement paste can be seen on the outer surface of the SAP

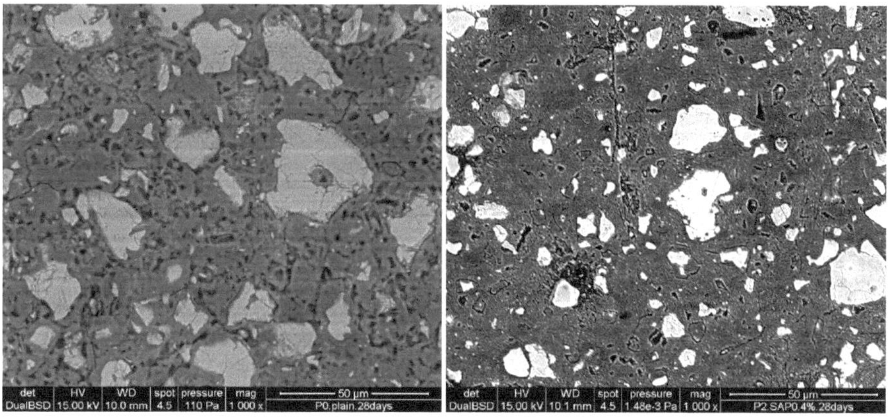

Fig. 6.7 Polished surfaces of plain cement paste (left) versus water-entrained SF-modified paste (right) as taken by BSE [7]

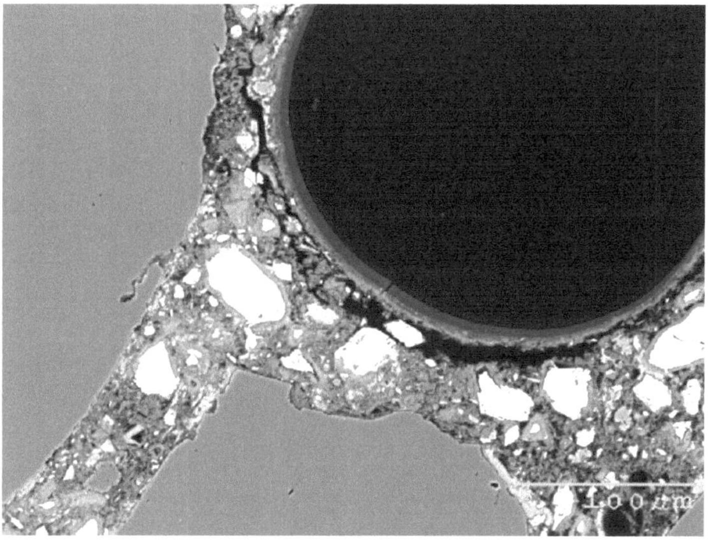

Fig. 6.8 Image of the interface of superabsorbent polymer specimen, view from SEM by using a BSE detector, Lam [4]

particle remnant, which has likely broken away from the cement matrix due to drying at the preparation leaving a gap.

Large voids, larger than 300 µm, also can be found in the mortar (Fig. 6.9). This is due to the non-uniform distribution of the SAP in the samples as discussed by Lam [4].

However, the interfacial transition zone (ITZ) appears differently when different experimental techniques are used and different sample preparation procedures are

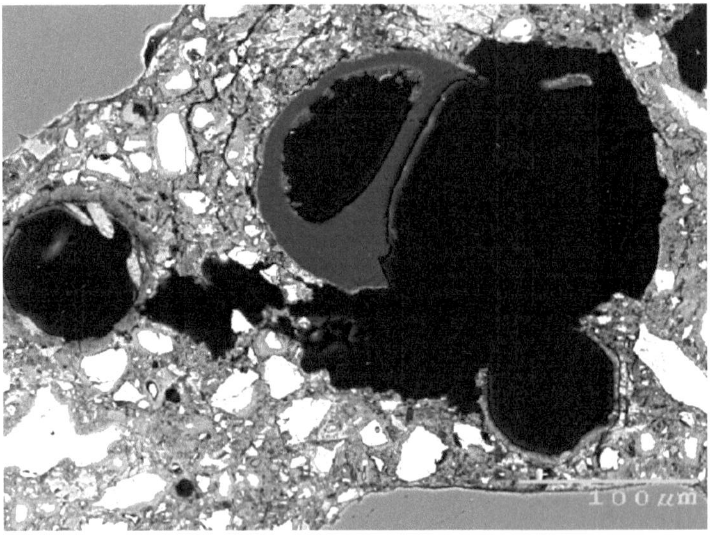

Fig. 6.9 Dispersion problem of SAP: Agglomeration phenomenon [4]

Fig. 6.10 GSE image of an SAP pore and surrounding concrete matrix at age 28 days [8]

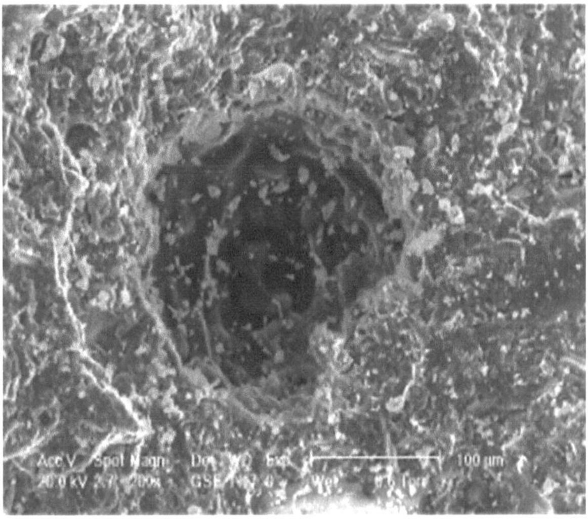

chosen. The ITZ observed by SEM in a gaseous secondary electron detector (GSE) mode shows totally different result comparing to a BSE detector mode. By using GSE, Mechtcherine et al [8] found that no differences were visible between the microstructure formed directly in the surrounding of SAP pores and the other matrix areas (Fig. 6.10). Similarly the investigation of the direct surroundings of SAP pores and other areas using X-ray diffraction showed no differences in the distribution of the elements.

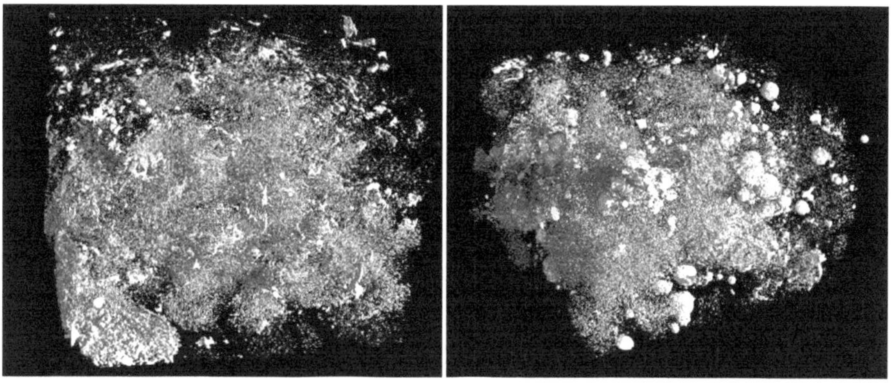

Fig. 6.11 Three-dimensional rendering of the pore structure of a plain mortar (left) and a mortar with SAP (right). The images were obtained by using micro-computer tomography. The samples are about 4–5 mm across [12]

Fig. 6.12 Micro-computer tomography image of pores entrained by using SAP [15]

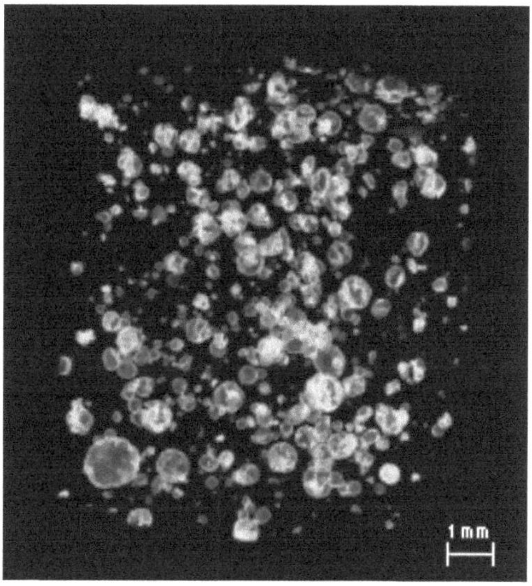

6.5 Structure of Voids Introduced by SAP in the Matrix

The structure of SAP introduced by SAP has a significant positive influence on the frost/thaw resistance in concrete [11]. Micro-computer tomography images of 3D distribution of SAP in mortars reported by Lura et al. [12] showed that in the mixture with SAP (Fig. 6.11, right image), a number of large, spherical pores are present. The pores introduced by SAP were evenly distributed over the concrete volume (Fig. 6.12) [8]. Fig. 6.13 gives a SEM image of an entrained water pore [13].

Fig. 6.13 Entrained pore in hardened UHPC [13]

It corresponds to the spherical shape of dry SAP particles, just like before their addition to the mixture. Due to the absorption of water, the pores are considerably larger than the dry SAP particles before their addition to concrete.

For 3D micro-computer tomography images shown in Fig. 6.11, the pore size and the pore size distribution by image analysis were performed in order to characterize quantitatively the features of the pore structure [10]. In the first step of the procedure, according to the gray level histogram, the black parts, which represent the voids and pore volume, were distinguished. In the next step, the volume of the pore in the image, was assessed by multiplying the number of pixels by the resolution of the image. Finally, the equivalent diameter of a sphere with a volume equal to that of the pore was calculated and the cumulative pore volume fraction against the equivalent pore diameter was obtained.

Cumulative pore (void) size distribution of the two mortars from Fig. 6.11 were computed and shown in Fig. 6.14. It is evident that the plain mortar shows a higher number of smaller pores of a few µm, which are the big capillary pores in the paste. Most of the porosity of the mortar with SAP is in the form of larger pores of tenths to hundreds of µm, which were the remains of the spherical SAP particles. The differential pore (void) size distribution of the two mortars (Fig. 6.14, right), which were calculated based on the 10 µm increments of the pore diameter, shows a peak at 20 µm for the plain mortar, while in the SAP mortar the peak is shifted to 150 µm. Notice that no pores smaller than about 10 µm could be detected with this type of microtomography.

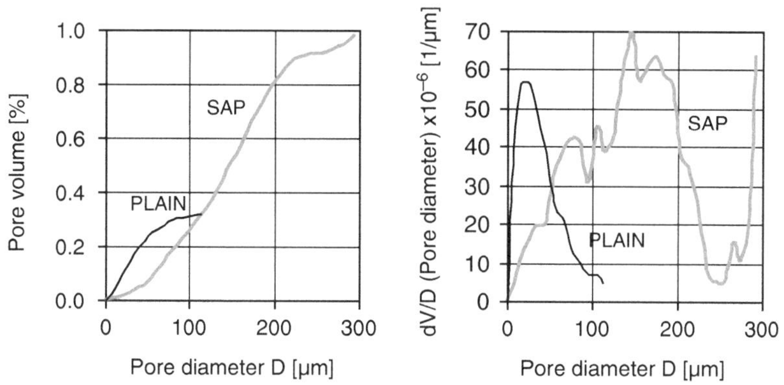

Fig. 6.14 Cumulative pore volume (left) and differential pore volume (right) as a function of pore diameter for a plain and a SAP-containing mortar [12]

6.6 Conclusion

SAP added in the concrete release the absorbed water and promote further hydration of cement particles. The degree of hydration of cement paste with SAP is higher after an age of 14 days compared to plain paste. The rate of the change of the degree of hydration depends on a couple of material parameters, such as the size and the amount of SAP in the mixtures, the original water/cement ratio and the procedure by which SAP was added in the concrete or mortar mixture.

The addition of SAP in the mixture increases the hydration degree of cement particle, this leads to a reduction of capillary porosity in the matrix. Micro-computer tomography images show that the voids introduced by SAP in the mixture are homogenously distributed.

References

[1] Jensen OM, Hansen PF (2001) Water-entrained cement-based materials I: Principles and theoretical background. Cem Conc Res 31:647–654

[2] Jensen OM, Hansen PF (2002) Water-entrained cement-based materials II: Experimental observations. Cem Concr Res 32:973–978

[3] Reinhardt W, Assmann A, Mönnig S (2008) Superabsorbent polymers (saps) – an admixture to increase the durability of concrete. 1st International Conference on Microstructure Related Durability of Cementitious Composites, 13–15 October 2008, Nanjing, China 313–322

[4] Lam H (2005) Effects of internal curing methods on restrained Shrinkage and permeability. portland cement association, Research & Development Information Serial No.2620

[5] Igarashi S, Watanabe (2006) A: Experimental study on prevention of autogenous deformation by internal curing using super-absorbent polymer particles. Int. RILEM Conference on Volume Changes of Hardening Concrete: Testing and Mitigation, RILEM Proceedings PRO 52, RILEM Publications S.A.R.L., 77–86

[6] Lura P, Durand F, Loukili A, Kovler K, Jensen OM (2006) Strength of cement pastes and mortars with superabsorbent polymers. Int. RILEM Conference on Volume Changes of Hardening Concrete: Testing and Mitigation, RILEM Proceedings PRO 52, RILEM Publications S.A.R.L., 117–126
[7] Esteves LP (2009) Internal curing in cement based materials, PhD Thesis, Aveiro University, Portugal
[8] Mechtcherine V, Dudziak L, Hempel S (2009) Mitigating early age shrinkage of concrete by using Super Absorbent Polymers (SAP). In: Creep, Shrinkage and Durability Mechanics of Concrete and Concrete Structures – CONCREEP-8, T. Tanabe et al. (eds.), Taylor & Francis Group, London, 847–853
[9] Mönnig S (2005) Water saturated superabsorbent polymers used in high strength concrete Otto Graf Journal 16:193–203
[10] Ye G (2003) Experimental study and numerical simulation of the development of the microstructure and permeability of cementitious materials. PhD thesis, Delft University of Technology, Delft
[11] Laustsen S, Hasholt MT, Jensen OM (2008) A new technology for air-entrainment of concrete. 1st International Conference on Microstructure Related Durability of Cementitious Composites, 13–15 October, Nanjing, China, 1223–1230
[12] Lura P, Ye G, Cnudde V, Jacobs P (2008) Preliminary results about 3d distribution of superabsorbent polymers in mortars. 1st International Conference on Microstructure Related Durability of Cementitious Composites, 13–15 October, Nanjing, China, 1341–1348
[13] Mechtcherine V, Dudziak L, Schulze J, Stähr H (2006) Internal curing by super absorbent polymers – Effects on material properties of self-compacting fibre-reinforced high performance concrete. Int. RILEM Conference on volume changes of hardening concrete: Testing and Mitigation, RILEM Proceedings PRO 52, RILEM Publications S.A.R.L., 87–96, 2006
[14] Dudziak L, Mechtcherine V (2010) Reducing the cracking potential of ultra-high performance concrete by using super absorbent polymers (SAP). In: Advances in Cement-based Materials, G. van Zijl and W.P. Boshoff (eds.), Taylor & Francis Group, London, 11–19
[15] Dudziak L, Mechtcherine V (2008) Mitigation of volume changes of ultra-high performance concrete (UHPC) by using super absorbent polymers. Proceeding of the second international symposium on ultra high performance concrete, E. Fehling et al. (eds.), Kassel University Press GmbH, 425–432

Chapter 7
Effects of Superabsorbent Polymers on Shrinkage of Concrete: Plastic, Autogenous, Drying

Viktor Mechtcherine and Lukasz Dudziak

Abstract This chapter deals prospectively with one of the main applications of SAP in concrete construction, the mitigation of the autogenous shrinkage of concrete. In particular, the effects of using SAP as an additive for internal curing in cementitious materials with a low water-to-cement ratio and a low-permeability microstructure are presented and discussed. Since the addition of SAP, often in conjunction with extra water, influences not only autogenous shrinkage but also other types of volumetric changes, this is addressed as well in this chapter. Furthermore, the development of stresses due to restrained autogenous shrinkage is also considered for concretes with and without internal curing.

7.1 Introduction

Volume changes in concrete due to shrinkage are major sources of eigenstresses and stresses due to the restraint in concrete structures; this leads in many instances to cracking of the concrete. Commonly four different types of shrinkage deformations are considered in this context: 1) *plastic shrinkage*, occurring during the first few hours after mixing the concrete when concrete still behaves as a formable mass; 2) *autogenous shrinkage* of hardening concrete as a result of the cement hydration process; 3) *drying shrinkage* caused by the loss of water to the surrounding atmosphere; and 4) *carbonation shrinkage* resulting from the carbonation process. This chapter focuses on the application of SAP as a means of internal curing to mitigate autogenous shrinkage and the resulting restraint stresses in their inception. In this context, *chemical shrinkage* is considered as one of the prerequisites for autogenous deformations. Furthermore, since the use of SAP and internal curing water leads to changes in the water's quantity and thermodynamic and kinetic availability, effects on other types of

V. Mechtcherine (✉) • L. Dudziak
Institute of Construction Materials, Technische Universität Dresden, Germany
e-mail: mechtcherine@tu-dresden.de

V. Mechtcherine and H.-W. Reinhardt (eds.), *Application of Superabsorbent Polymers (SAP) in Concrete Construction*, RILEM State of the Art Reports 2,
DOI 10.1007/978-94-007-2733-5_7, © RILEM 2012

shrinkage deformations can be expected and as such are addressed in this chapter as well. In view of the information's present unavailability with reference to the effect of internal curing on carbonation shrinkage, this subject is not discussed here.

7.2 Plastic Shrinkage

7.2.1 Mechanisms of plastic shrinkage

Concrete at a very early age, i.e., after mixing and casting but before hardening, is susceptible to specific volumetric change: a phenomenon known as plastic shrinkage. The term reflects the concrete's plastic state (fluid or semi-fluid), when it exhibits properties of a drying suspension consisting of densely-packed solid particles distributed in a liquid phase.

The underlying reason for plastic shrinkage is the rapid loss of water from the concrete surface by evaporation. Thus, insufficient curing and a lack of protection of the concrete surfaces exposed to unfavourable ambient conditions (strong wind, high temperature, and/or low humidity) are common causes of marked plastic shrinkage in concrete. Irrespective of the actual cause, plastic shrinkage takes place because the magnitude of water losses is greater than the amount of water available to replenish the surface moisture lost. Accordingly, a direct consequence of this disproportion is the physical process of capillary pressure build-up [1], which relates to the connectivity between the solid components of a drying suspension [2, 3].

As long as the water remaining in the system covers all the superficial solid particles, there is no pressure difference between the water and surrounding air, and, therefore, no capillary pressure is generated. However, once the water film covering concrete surface vanishes (exposing individual solid particles), water menisci begin developing. Adhesive forces and surface tension give rise to capillary pressure, which starts to increase in inverse proportion to the radii of the forming menisci. As a result, contracting interparticular forces induce both settlement and shrinkage deformation along the exposed concrete surface [4]. Eventually, at particular points in time, the cavities can no longer be bridged by the water menisci. At this moment air penetrates the system, causing local breakdowns in hydrostatic pressure. Figure 7.1 from [2] gives an illustration of this process. In the presence of movement restraint, visible cracking will eventually take place as a result of plastic shrinkage.

7.2.2 Measuring plastic shrinkage

Various problems associated with the accurate assessment of early age volume changes have led to the development of an array of different measuring protocols. It is known that plastic shrinkage may be coupled to changes in early age deformations, for example, one-dimensional settlement and three-dimensional capillary shrinkage. For a better understanding of the physical processes behind these volumetric changes, information

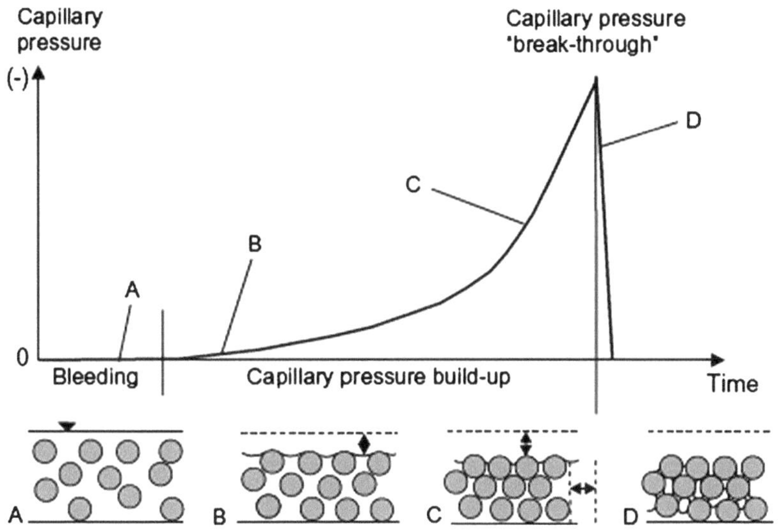

Fig. 7.1 Visualization of capillary pressure build-up, according to Slowik et al. [2]. The particular stages of the physical process leading to plastic shrinkage are indicated by the letters A to D

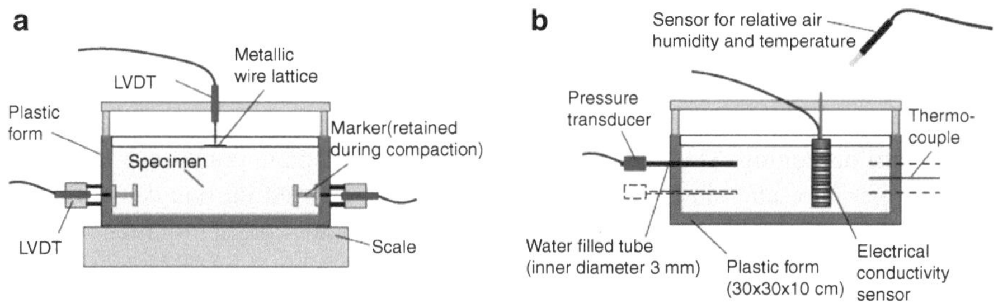

Fig. 7.2 Experimental set-up developed by Slowik et al. [2]. One mould (**a**) enables the measurement of horizontal shrinkage and settlement. The other one (**b**) is used for recording the changes in capillary pressure and electrical conductivity. Additionally, changes in ambient and concrete temperatures and in relative humidity during testing are traced with the second mould

on the development of capillary pressure and electrical conductivity has to be acquired as well. Furthermore, general information on losses of mass and temperature development are needed. An example of the equipment required to determine the variables mentioned above is the one proposed by Slowik et al. [2] and illustrated in Figure 7.2.

One of the important issues in this type of testing is the placement of the sensors to measure the plastic deformations. It is generally known that the evaporation of free water leads to differential shrinkage at the top surface. Particularly when concrete is exposed to extreme temperatures, differential thermal dilation may be of great importance as well. Accordingly, to measure plastic deformation at the location of the highest tensile stress, a reasonable solution is to place the measuring sensors at the top surface. The test setup fulfilling these requirements was proposed by Kratz et al. [5], and is shown in Figure 7.3.

Fig. 7.3 Test setup used by Kratz et al. [5] for plastic shrinkage measurements. The plastic deformation is traced using a LVDT near the surface. Developing ambient and concrete temperatures are recorded simultaneously

7.2.3 Effect of SAP addition on plastic shrinkage

At present only limited information exists on the particular effects caused by the use of SAP in cement-based materials with regard to plastic shrinkage and related parameters. An experimental investigation by Dudziak and Mechtcherine [6] seeking to determine the plastic behaviour of fresh cement pastes was conducted for 24 hours until hardening. The test, executed under harsh ambient conditions (constant high temperature and simulated wind blow), was conducted on one cement paste made with a water-to-cement ratio w/c = 0.3, and on another paste modified by SAP (0.6% by mass of cement) with some extra added water ($w_e/c=0.087$). Covalently cross-linked acrylamide/acrylic acid copolymers of spherical shape and an average particle size of 150 μm in the dry state were used as a water absorber. Both cement pastes contained equivalent amounts of superplasticizer and reached equal workability, i.e., the slump flow ranged from 190 to 195 mm. This was concluded to be proof that the entire extra added water was absorbed by SAP during the mixing process. Using a setup similar to that of Slowik et al. [2], the authors found that plastic deformation and capillary pressure of the cement paste were clearly reduced due to the addition of SAP and extra water while the settlement deformation increased compared to the control mixture, cf. Figure 7.4.

Particular changes observed for the paste with SAP and some extra water coincided with other effects, e.g., retardation of setting for the paste with SAP and extra water. As an explanation given to the less pronounced settlement of the cement paste without SAP, Dudziak and Mechtcherine [6] pointed to two phenomena acting simultaneously: (i) earlier and higher rise of temperature due to heat of hydration; and (ii) high positive autogenous deformations obtained during the first hour or two after casting. Both of these phenomena were observed for the control paste free of SAP but were not apparent for the paste enriched with SAP and extra water. In the

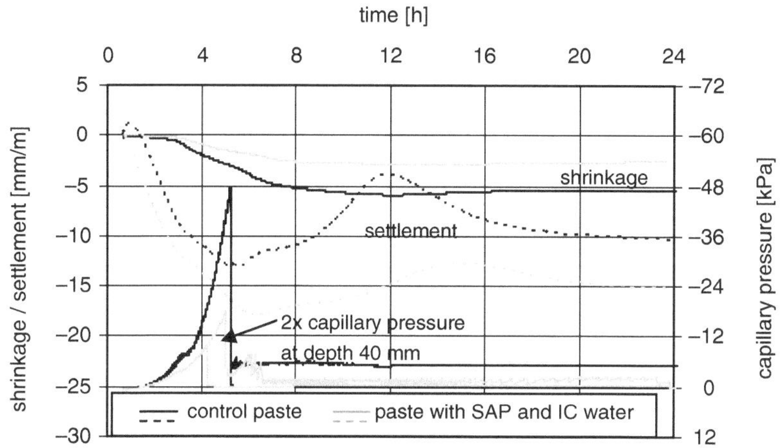

Fig. 7.4 Evolution of plastic deformation, settlement, and capillary pressure build-up over time for cement pastes with and without addition of SAP and extra water [6]

continuation of this study, the capillary pressure, a variable reflecting to some extent the cracking propensity of the fresh material, was found to be considerably lower for the paste with SAP and extra water in comparison to the reference mixture [7].

Kratz et al. [5] studied the effect of SAP's internal curing on the plastic behaviour of concretes made with different water-to-cement ratios. The mixtures were purely cement-based, containing 300-450 kg/m³ of binder and having w/c values in the range of 0.4 to 0.6. River gravel with a maximum grain size of 8 or 16 mm was used as coarse aggregate, and natural river sand as the fine aggregate. The total content of aggregate varied between 73 and 79% by volume. The concrete compositions were varied by the presence of SAP (0%, 0.3% or 0.6% by weight of cement), the amount of the extra water (w_e/c = 0, 0.03, and 0.06, respectively) and the addition of a superplasticizer (from 0.2 to 4.25% by mass of cement, dependant on the w/c ratio of the mix and the effectiveness of the superplasticizer). The SAP used was a powder of covalently cross-linked acrylamide/acrylic acid copolymers of a spherical shape and an average size of approximately 200 μm in the dry state. All mixtures had nearly identical workability as measured by the slump test. Plastic deformation at the top layer and loss of mass were ascertained using the test setup shown in Figure 7.3. Accordingly every mix studied was cast into a mould with a length of 700 mm and a cross-section of 150 mm x 150 mm. The testing conditions were standard laboratory atmospheric conditions (temperature 20°C, relative humidity 65%) or exposure to uniform air flow (ambient or hot air).

As a general tendency the values of plastic deformation were found considerably lower for various SAP and extra water contents enriching the concrete mixtures under study. In fact, a common behaviour observed for SAP-mixes was either expansion or low plastic deformation, whereas SAP-free mixes underwent pronounced plastic shrinkage. The results for the mixes with SAP but without additional water

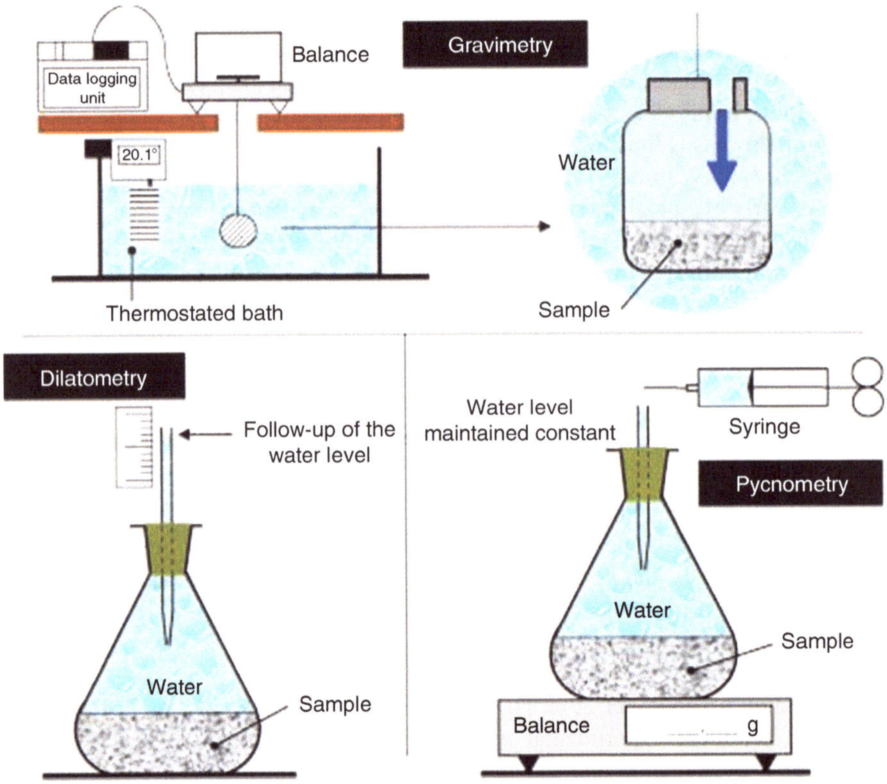

Fig. 7.5 Chemical shrinkage measurement methods presented in [10] based on [17]

paste (0.4% SAP by weight of cement); and a paste containing silica fume (15% SF by weight of cement). The basic water-to-cement ratio (corresponding to w/c_{eff}) was 0.30 for all mixtures. Unlike the control cement paste, the SAP-enriched paste contained some extra added curing water (5% by weight of cement, i.e., $(w/c)_{IC}$=0.05). A naphthalene-based superplasticizer was added to the paste containing silica fume in the amount of 1% by weight of cement. In the test protocol used, approximately 8 g of freshly mixed cement paste was cast in the bottom of a small glass jar of diameter 25 mm and height 60 mm. After the 10 mm-thick sample layer was covered by 1 mm of water, a metal vial encasing the specimen and glass container were submerged into a temperature controlled bath. Figure 7.6 shows the measured chemical shrinkage deformations in ml/g cement for the three paste systems.

The highest chemical shrinkage observed in all mixtures up to the age of 10 days was obtained for the SAP-enriched paste and was clearly driven by the highest rate of development in the first 2 days of hydration. The change attributed to SAP was suggested to be a result of the higher moisture content present as entrained water in the paste, which in turn favoured the movement of water from the external source. It was, however, pointed out by the author that the results obtained were

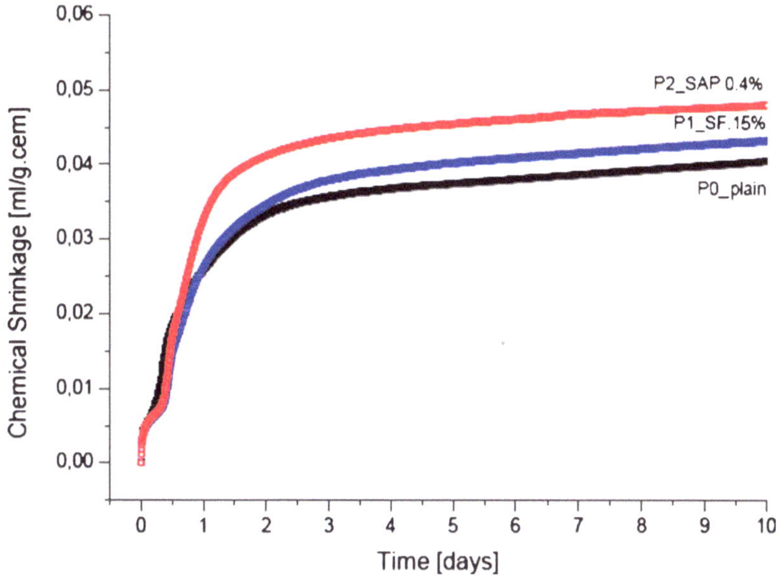

Fig. 7.6 Effect of SAP and silica fume on chemical shrinkage of cement paste. The basic water-to-cement ratio was 0.30 [19]

somewhat questionable since the chemical shrinkage recorded for the plain cement paste was below 0.04 ml/g of cement during the first 7 days. This value was considered to be noticeably lower than typically-recorded values for systems containing pure cement paste (approx. 0.06 ml/g of cement). The author suggested that the main reason for this difference was insufficient water uptake by pastes with low water-to-cement ratios, i.e., the ability of water to penetrate into the sample might be reduced due to a dense pore structure. Note that Lura [16] reached a nearly identical value of chemical shrinkage for a cement paste made with w/c of 0.3, i.e., approximately 0.04 ml per gram of cement after 7 days using a 5 mm thick specimen (about 5 g) and a similar experimental protocol.

7.4 Autogenous Shrinkage

7.4.1 Mechanisms of autogenous shrinkage

Autogenous shrinkage was first recognised and described more than sixty years ago by Lyman [20] when a decrease in the volume of concrete was observed without any notable change in its mass or temperature. However, since autogenous shrinkage of ordinary concrete is considerably smaller than drying shrinkage, not much attention was dedicated to this subject until the beginning of application of high-performance concrete in construction, where it was observed that HPC was prone to autogenous-shrinkage-caused cracking under restraint. In the last two decades the phenomenon of autogenous shrinkage has been studied with increasing intensity. Now, formulas for predicting autogenous shrinkage are included in a number of design codes for concrete structures (DIN 1045-1 [21], EN1992-1-1 [22], JSCE Design Code 2002 [23]).

Autogenous shrinkage is typically defined as the bulk deformation of a closed (sealed), isothermal, cementitious material system not subject to external forces [24]. Before the setting of concrete, autogenous shrinkage has been regarded as equal to chemical shrinkage, i.e., the internally occurring and the externally measurable volume changes are alike. It is attributed to the fact that a cementitious material in a liquid-like state is not able to sustain the internal voids created by chemical shrinkage. After the formation of a solid skeleton of hydration products, however, the bulk, macroscopically observed deformations of the entire concrete system, are clearly smaller than the changes in the internal volume of the material. Under microscopic observation, the formation of a large number of fine pores in the cement paste can be regarded as the manifestation of significant volume changes due to hydration. The macroscopically measured reduction in concrete specimen volume (autogenous shrinkage) can be, therefore, only a fraction of the real material shrinkage (chemical shrinkage).

To what extent chemical shrinkage is transformed into autogenous shrinkage depends obviously on the water content of the system and the pore size distribution resulting from hydration. As mentioned above, autogenous shrinkage is relatively small in comparison to drying shrinkage in the case of ordinary concrete. Mainly for reasons of workability, more water is used in the production of such concrete than is needed for the complete hydration of cement. Thus, pores forming due to chemical shrinkage are either filled with water, or at least the air in the concrete pores has a relative humidity near 100%. This means that during the entire hydration process no significant self-desiccation occurs and the thermodynamic equilibrium of the system is undisturbed to a great extent. Similar considerations can be offered with regard to the disjoining pressure between the particles of the C-S-H gel. Hence, there is no pronounced driving force for autogenous deformations in typical, normal-strength concrete.

On the contrary, in high-strength concrete/high-performance concrete (HSC/HPC) the amount of water is insufficient to achieve complete hydration of the cement due to the low water-cement ratio (w/c clearly below 0.4). This shortage of free water results in a pronounced decrease in the relative humidity within the pore system. Two mechanisms are likely to be responsible for the development of such shrinkage deformations from self-desiccation: 1) a decrease in the disjoining pressure between the particles of the C-S-H gel, and 2) a decrease in the menisci radii of the pore water, which increases the tension both within the pore water and at its surface. As a result of the second phenomenon, compressive stresses develop simultaneously in the solid skeleton of the cement system to restore equilibrium by compensating the tensile stresses in the fluid phase. This results in a volume reduction of the entire system, i.e., autogenous shrinkage. As of today it is not known to which extent the first or the second mechanism dominates the material behaviour observed. Further mechanisms are possibly involved as well.

Despite the low water-to-cement ratio, HSC/HPC mixtures usually contain a superplasticizer and silica fume, both leading to an increase in autogenous shrinkage. It is believed that silica fume contributes to autogenous shrinkage because it leads to a finer pore structure and thus increases the capillary stress. Moreover, the pozzolanic reaction of silica fume involves chemical shrinkage, thus further contributing to the bulk shrinkage of SF-enriched concrete.

7.4.2 Application of SAP for mitigation of autogenous shrinkage

Conventional methods of curing concrete cannot contribute substantially to the mitigation of autogenous shrinkage of concrete with a low water-to-cement ratio, even if intensive wet curing is applied. In contrast to drying shrinkage, which occurs due to water loss at the surface of concrete members, autogenous shrinkage occurs over the entire volume of the concrete member, and consequently a surface treatment cannot suffice. Furthermore, since the microstructure of typical HSC/HPC is very dense even at early ages, it does not allow a sufficiently rapid transport of curing water into the interior of concrete members, especially if they are thick.

With that in mind, the use of internal curing (IC) by adding materials with a high water storing capacity to the mixture has been proposed, which would supply water to the surrounding matrix as soon as self-desiccation occurs. Since the 1990s, several researchers have investigated the effect of the addition of a portion of water-saturated lightweight aggregate (LWA) on the properties of HSC [24, 25, 26]. Later, other materials were proposed as IC-agents, of which superabsorbent polymers (SAP) appear to be the most appropriate for use as a new water-regulating admixture in concrete. Following the work of Jensen and Hansen [27, 28], they and other researchers tested various types of SAP and demonstrated that certain of them possess a very pronounced ability to mitigate autogenous shrinkage in HSC [29, 30].

The advantage of SAP is that they can be in principle engineered for the special purpose of internal curing by designing the necessary particle size and shape, water absorption capacity, and other properties. Only small amounts of SAP plus some additional water for internal curing are added directly to the fresh mixture.

7.4.3 Measuring the reduction of autogenous shrinkage due to use of SAP

To quantify the efficiency of internal curing through the use of SAP, reliable measurement techniques for autogenous shrinkage are required. This issue was recently addressed by two RILEM Technical Committees: TC-181 EAS "Early Age Shrinkage

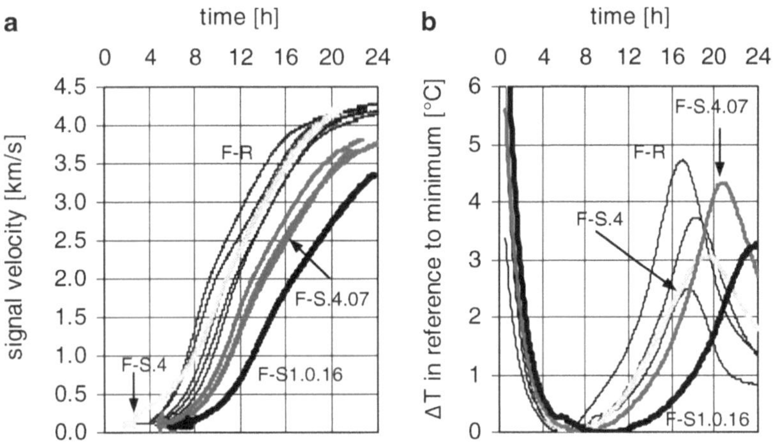

Fig. 7.9 Effect of SAP and extra water addition on the results of: (**a**) ultrasonic tests and (**b**) temperature measurements [40]

shrinkage, other researchers such as Mechtcherine et al. [29, 35] used the final set time as a starting point for the measurement of autogenous shrinkage.

Such t_0 measurements using the testing procedure according to DIN 480-2 (Measurement of change in pin penetration in mortar over time [36]) were performed by Dudziak and Mechtcherine [40] on a finely grained, ultra-high-performance concrete (w/b=0.19) with and without SAP. They showed that the setting time can increase from 8.6 hours (reference mixture) to approximately 10.4 hours (addition of 0.4% SAP by cement weight and extra water corresponding to $(w/c)_{IC} = 0.07$) and even to 12.5 hours (addition of 1.0% SAP by cement weight and extra water corresponding to $(w/c)_{IC} = 0.16$). The corresponding measurements of temperature and ultra-sound velocity indicated similar tendencies, cf. Figure 7.9 (F-R was the reference mix without SAP, other mixes contained SAP and, apart from F-S.4, also had some extra water). The particular mixtures considered in the study and their nomenclature are addressed in the following sections of this chapter.

Regardless of the applied technique for measuring t_0, or even applied combinations of different techniques, uncertainty remains concerning the accuracy in the selection of the particular t_0 value. This makes the interpretation of the autogenous shrinkage data complicated, especially due to the fact that great changes in the autogenous deformation behaviour of concrete occur around this time t_0.

The measurements by Ribeiro [41] showed that the values measured in corrugated tubes are on the same order of magnitude as the chemical shrinkage measurements obtained using the ASTM C1608 Method A [42]. From this finding, Ribeiro concluded that the initial slope of the curves from the tests with corrugated tubes corresponds to the chemical shrinkage of concrete in the fluid state. Furthermore, he pointed out that as far the slope of the curve does not change noticeably, the mechanism leading to the deformations should be the same, i.e., chemically-based. A transition to restrained autogenous shrinkage of a solid skeleton, which would induce tensile stresses in the system, should be marked by a clear change in the slope of the curve according to [41]. Further research would be needed to substantiate this assumption.

Finally, the effect of the temperature during testing has to be addressed briefly here. It should be considered that even very moderate temperature changes can induce thermal deformations affecting the results of shrinkage measurements. For instance, the temperature development in concrete as a result of cement hydration can differ due to the application of internal curing, i.e., a slight shift in the temperature curve toward a higher concrete age combined with a lower temperature peak, cf. Figure 7.9b. Therefore, when interpreting the data obtained, possibly a distinction should be made between thermally induced and shrinkage deformations in order to attain a more comprehensive view of the existing interactions.

7.4.4 Effect of SAP addition on autogenous shrinkage

Since internal curing is meant to mitigate the volume reduction of hardening and hardened cement paste, the effect of SAP on autogenous shrinkage can be best demonstrated on this component of concrete. It is common knowledge that the shrinkage of concrete (autogenous or otherwise) is generally smaller in comparison to that of cement paste. Considering the possible differences between the effects of SAP addition to cement paste versus concrete (or mortar), two opposing issues seem to be of relevance: 1) The distribution of SAP particles over the cement paste volume might be more homogenous in the case of concrete due to the generally higher mixing intensity used, and 2) the presence of aggregates and fibres in concrete may put some additional mechanical loading on SAP in fresh concrete, which could lead to the destruction of some unstable SAP particles. In the subsequent subchapters the results demonstrated in the literature are presented separately for cement paste, mortar, and concrete.

7.4.4.1 Cement paste

The effect of SAP on the reduction of autogenous shrinkage was studied by Jensen and Hansen [28]. They measured autogenous deformations of hydrating cement pastes containing different amounts of SAP (suspension-polymerized having spherical particles with an average particle size of $\approx 200\ \mu m$) under isothermal conditions, i.e., ambient temperature of 20°C. The autogenous deformations of cement paste specimens made using a basic w/c = 0.3 were measured by a dilatometer with corrugated polyethylene moulds. For easy comparison, each set of deformation measurements was adjusted to zero at the time of setting, which occurred approximately 9 hours after water addition to cement. Figure 7.10 shows that without SAP addition, a significant shrinkage of approximately 3700 microstrains developed in the cement paste over 3 weeks of sealed curing. The authors found that additions of 0.3% and 0.6% SAP by weight of cement led to a successful mitigation of shrinkage and even induced some expansion.

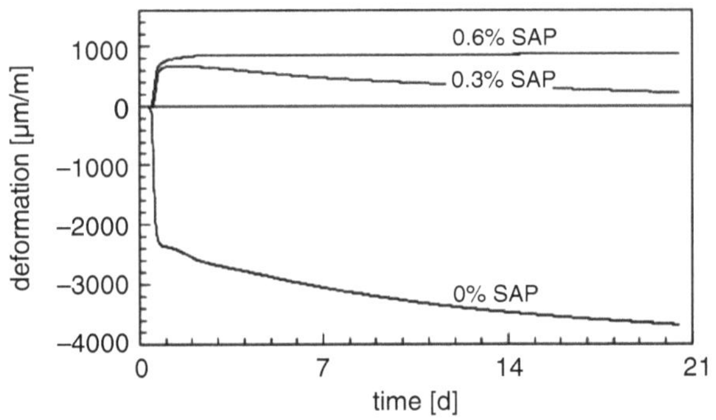

Fig. 7.10 Autogenous deformations measured from setting of cement pastes with different amounts of SAP (Type A) and, consequently, different amounts of entrained water [28]. SAP additions are given by mass of cement, where 0.6% corresponds to an entrained w/c of 0.075. Basic w/c was 0.3 for all mixes, and the temperature was constant at 20°C

Another experimental investigation of cement paste was performed by Igarashi and Watanabe [30]. Figure 7.11 shows the autogenous shrinkage deformations of ordinary Portland cement pastes with water/cement ratios of 0.25 and 0.33, respectively. The addition of SAP (0.35% by mass of cement, $w_{IC}/c = 0.045$) drastically reduced autogenous shrinkage at a water/cement ratio of 0.25. When the doubled amounts of SAP and internal curing water were incorporated into the pastes, autogenous shrinkage was completely prevented. However, autogenous shrinkage was not prevented completely at a higher water/cement ratio of 0.33, even when a very high amount of SAP was incorporated into the paste. It should be noted that no increase in shrinkage was observed after the first 24 hours in the specimens with SAP for the water/cement ratio of 0.33. The reasons of such a pronounced difference in the results for pastes with w/c ratio of 0.25 and 0.33 are still to be investigated further.

In the same study [30] autogenous shrinkage of a paste containing silica fume with a water/binder ratio of 0.25 was tested. The authors reported the reduction of autogenous shrinkage to approximately one third by the addition of 0.35% SAP by weight of cement and a corresponding amount of extra water (cf. Figure 7.12). Although these results were similar to those for the cement paste without silica fume, the cement paste with SF and 0.70% SAP still exhibited slight shrinkage at very early ages. This is in contrast to the corresponding ordinary cement paste, for which this amount of SAP completely prohibited autogenous shrinkage (Fig. 7.11a).

The effect of SAP-particle size and content was studied by Lura et al. [43]. The authors used a suspension-polymerized, covalently cross-linked acrylamide/acrylic acid copolymer as SAP. The spherical particles have diameters varying from 50 to 250 µm in the dry state. The size of the expanded SAP particles in the cement pastes was 2-3 times greater due to pore fluid absorption. During mixing of the cement pastes, the SAP absorbed approximately 12.5 g of pore fluid per gram of SAP. Additional mixing water was added to the mixtures with SAP in an amount sufficient

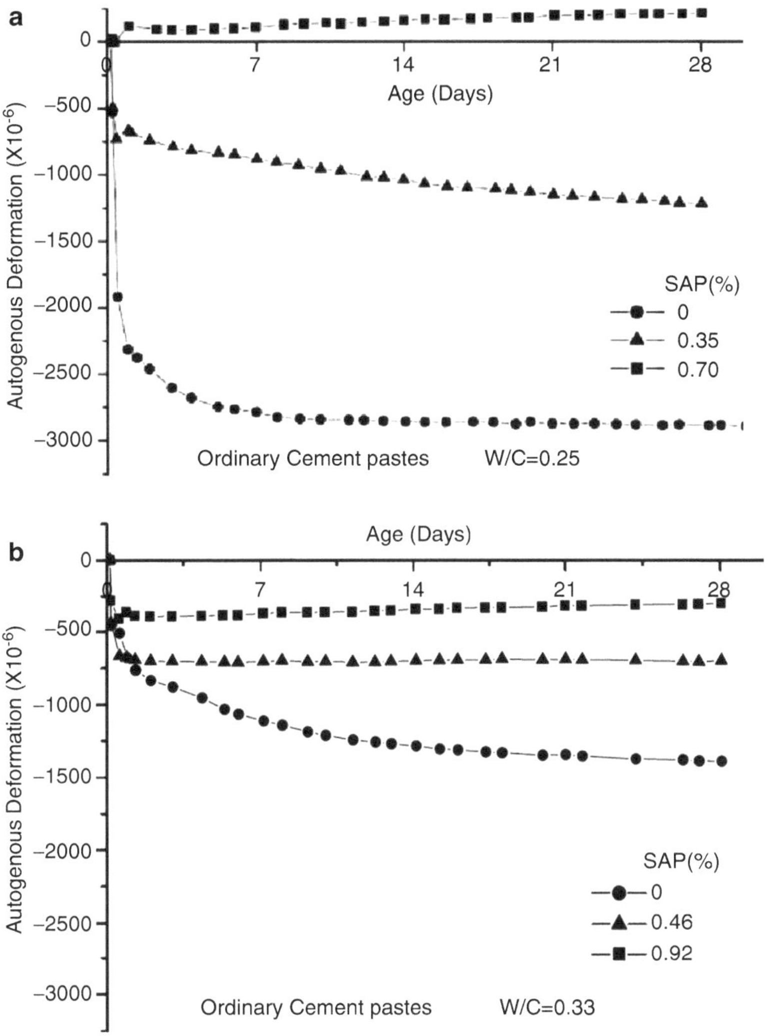

Fig. 7.11 Autogenous deformations of ordinary Portland cement pastes with different amounts of SAP: (**a**) for w/c=0.25, and (**b**) for w/c=0.33 [30]

to saturate the SAP particles. This water amount for internal curing corresponded to w_{IC}/c=0.037 and 0.075, respectively, for the pastes with SAP addition of 0.3% and 0.6% by weight of cement. The authors showed that in contrast to the usually expected enhancement of IC efficiency due to the better particle distribution in the pastes with smaller-size particles, the overall IC efficiency was somewhat lower than that of the mixes with larger SAP particles (Figure 7.13 a & b). The particle size influenced mainly the initial expansion observed during the first day of measurements. The authors suggested that the small-sized SAP particles released IC water less promptly

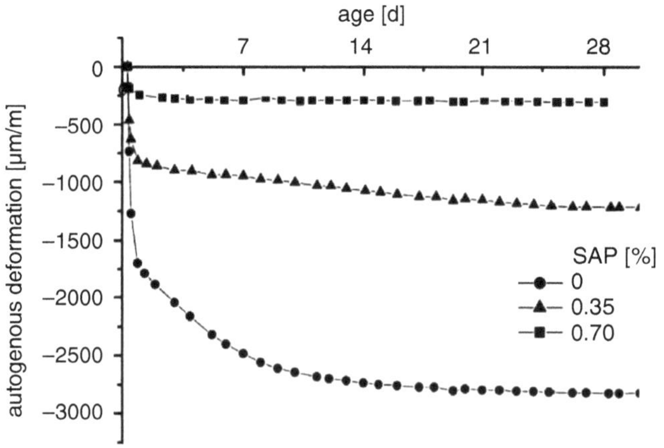

Fig. 7.12 Autogenous deformations of cement pastes containing silica fume and different amounts of SAP, w/c=0.25, after [30]

at early ages, and that the bulk of the SAP acted as the main water reservoir. With the doubled amount of SAP (0.6% SAP) and extra water, which were sufficient to avoid self-desiccation completely in the first two weeks of observation, the effect of particle size became less pronounced.

With regard to the water absorption capacity of SAP with different particle sizes, Esteves [19] found that smaller particles reached saturation much more rapidly but absorbed less water in comparison to larger particles of the same chemical composition and production process.

7.4.4.2 Mortar

Only a few investigations of mortar are known. Geiker et al. [44] compared two different methods of internal water supply in cement/silica fume mortars made with w/b=0.35: (a) replacement of a portion (8% and 20%) of the sand by water saturated fine LWA; and (b) the addition of SAP (0.4% by mass of cement). The results presented in Figure 7.14 demonstrated that addition of a small amount of SAP is as efficient with regard to mitigating autogenous shrinkage as the replacement of a considerable amount of normal weight aggregate by LWA.

7.4.4.3 Concrete

Igarashi and Watanabe studied the effect of IC by means of SAP on autogenous deformations of concrete made with w/c = 0.25 [30]. The mixture contained silica fume (10% by mass of cement), river gravel (maximum size of 10 mm) as coarse aggregate, natural river sand as fine aggregate, and a polycarboxylic acid type

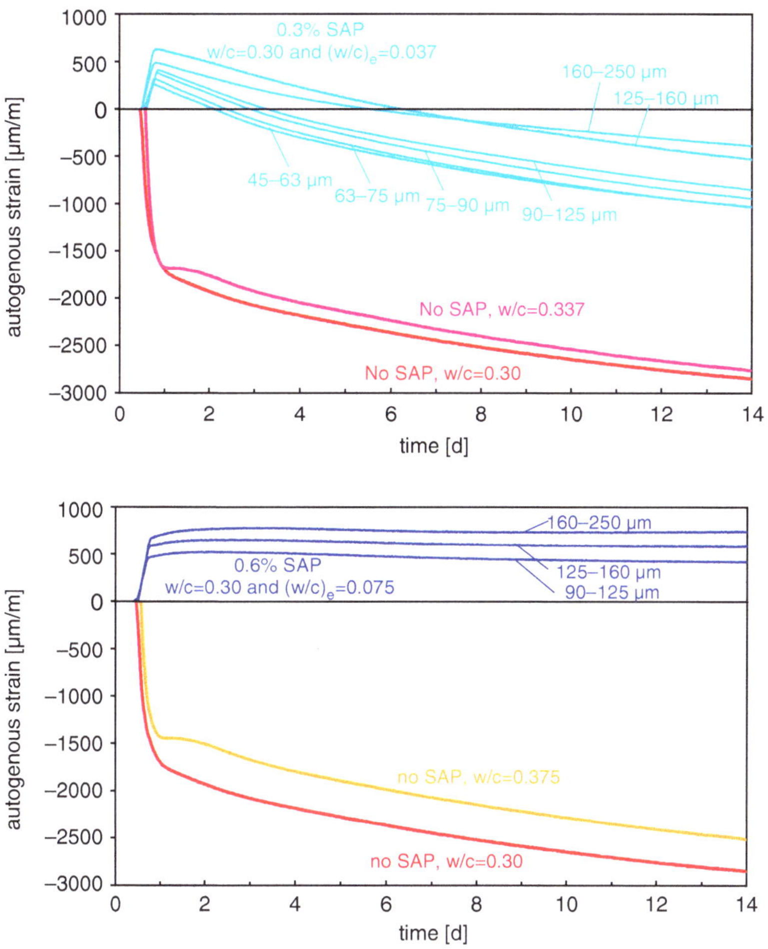

Fig. 7.13 Autogenous strains at 20°C measured on cement pastes with w/c=0.30 and 20% silica fume addition, with no SAP, 0.3% SAP (top) and 0.6% SAP (bottom) by weight of cement (further results in ref. [43])

superplasticizer. The SAP used was a powder of acrylamide/acrylic acid copolymer. Trends similar to those found for cement pastes and mortars were observed (cf. Figure 7.15) as far as SAP shrinkage reduction effectiveness is concerned; however, the autogenous deformations of the reference concretes were considerably lower in comparison to those measured for the reference cement pastes. A moderate addition of SAP and IC water (0.35% SAP by mass of cement, and $w_{IC}/c=0.05$ which was sufficient to saturate SAP particles) led to a pronounced reduction of autogenous shrinkage. When the amount of SAP was doubled, the concrete exhibited a small expansion during the initial 24 hours, which was then maintained up to an age of 7 days. The autogenous shrinkage of the concrete with

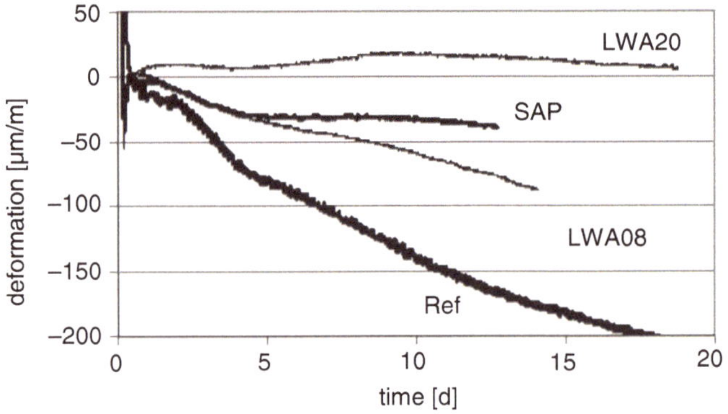

Fig. 7.14 Autogenous deformation vs. time measured for mortars made with IC agents (LWA, 8% and 20% of sand replacement, and SAP) and without internal curing (Ref). The specimens were sealed and tested at a constant temperature of 30ºC [44]

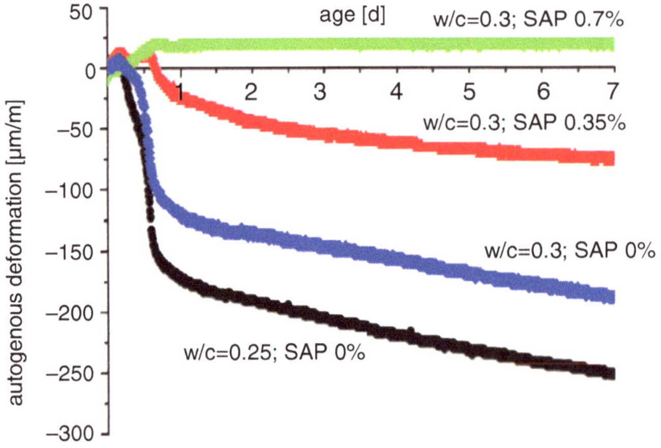

Fig. 7.15 Autogenous deformation vs. time for concretes with and without addition of SAP [30]

a higher water-to-cement ratio of 0.30 was smaller than that of the reference concrete with the water/cement ratio of 0.25, but by far larger than that of the concrete with the same water-to-cement ratio (w/c = 0.3) and SAP addition. This clearly shows that the shrinkage reduction in concrete containing SAP resulted from the effective internal curing with the SAP and not from the possible increase in effective water-to-cement ratio which could have been due to a smaller than expected water absorption by SAP.

Mechtcherine et al. [29, 45] developed and investigated a self-compacting fibre-reinforced high-performance concrete with internal curing, which was subsequently applied in building the first known structure with SAP used as an internal curing agent, cf. Chapter 10. This concrete had a Portland cement content of 800 kg/m^3,

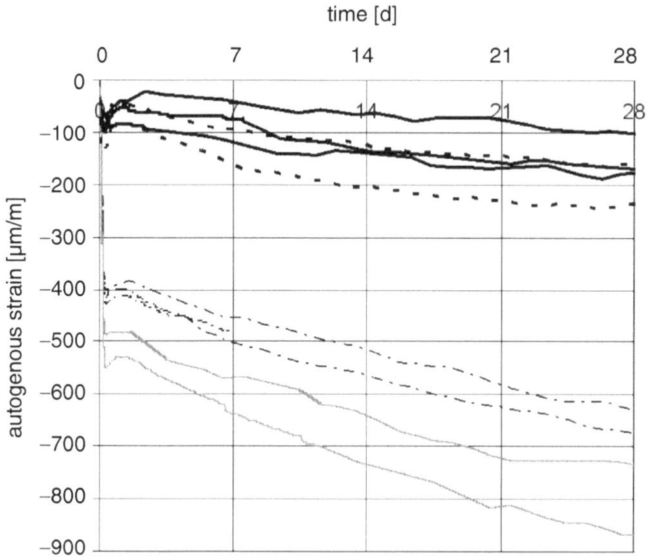

Fig. 7.16 Autogenous shrinkage during the first 28 days after setting for concrete mixtures with and without SAP [29]

a silica fume content of 120 kg/m³, and a water-to-cement ratio of 0.24. Fine quartz sand (0.125 - 0.5 mm), crushed basalt sand (0 - 2 mm) and coarse crushed basalt split (2 - 5 mm) were used as aggregates. Furthermore, a polycarboxilat-ether based superplasticizer and 144 kg/m³ of fine steel fibres (length 6 mm, diameter 0.15 mm) were used in the mix. Covalently cross linked acrylamide/acrylic acid copolymers were used as SAP. The dry powder consisted of suspension-polymerised spherical particles with an average size of approx. 200 µm. The amount of SAP was 0.4% by mass of cement, with added IC water $w_{IN}/c = 0.05$.

Figure 7.16 shows representative autogenous strain-time curves obtained for the compositions with and without SAP during the first 28 days (two to three curves are shown for each concrete composition). These curves were zeroed at the end of concrete setting so that only potential stress-inducing strains were considered. The mixtures with SAP addition (Ref-SAP, M17) showed a dramatic decrease in autogenous shrinkage in comparison to the corresponding mixtures without SAP (Ref, M12). For example, at a concrete age of 3 days, the autogenous shrinkage of the reference concrete Ref-SAP was on average about 50 µm/m while the corresponding value for the reference concrete without SAP was above 400 µm/m.

The effect of internal curing on the autogenous shrinkage of a fine-grained, ultra-high performance concrete (UHPC) was studied by Mechtcherine et al. [35, 46]. The reference mixture contained 853 kg/m³ of Portland cement and 138.5 kg/m³ of silica fume. Quartz powder and quartz sand were used as aggregates. The water-to-binder ratio was 0.19, and 30 kg/m³ of polycarboxilate-ether based superplasticizer

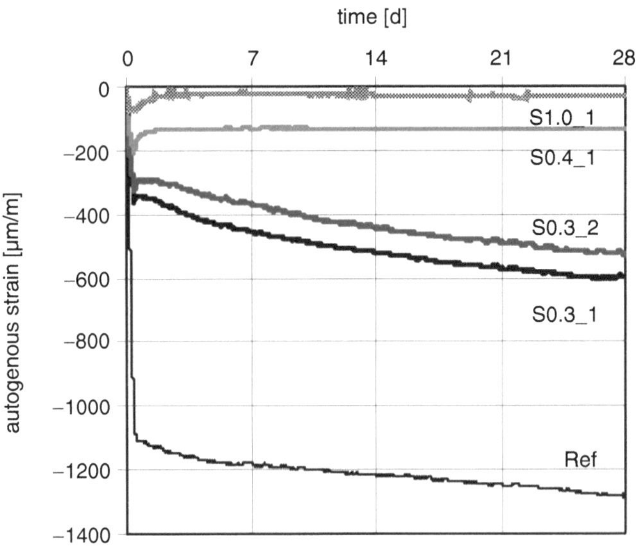

Fig. 7.17 Effect of the different amounts of SAP and internal curing water on autogenous shrinkage of UHPC measured on corrugated tubes starting after the final set [46]

were added to the mix to achieve the desired workability. The variable parameters were the amount of internal curing agent (SAP) and the amount of additional water supplied for internal curing. The SAP used were suspension-polymerized spherical particles with an average particle size of 150 µm, added in dry condition during mixing. They consisted of covalently cross-linked acrylamide/acrylic acid copolymers. In general, all the mixtures containing SAP and extra water showed pronounced reductions in shrinkage deformations. These reductions were particularly dramatic at a very early age, in fact in the first twelve hours after the final set, cf. Figure 7.17. The addition of 0.3% SAP by mass of cement (mixtures S0.3_1 and S0.3_2) plus extra water (corresponding $w_{IC}/c=0.04$ and $w_{IC}/c=0.05$, respectively) resulted in a decrease in autogenous shrinkage from approximately 1100 µm/m (mixture Ref) to approximately 300 µm/m in the first day of measurement. With increasing amounts of SAP and IC water (mixture S0.4_1: 0.4% SAP and $w_{IC}/c=0.07$; mixture S1.0_1: 1.0% SAP and $w_{IC}/c=0.16$), an even more considerable reduction in autogenous shrinkage deformations was recorded (cf. Figure 7.17), making autogenous shrinkage of these UHPC mixtures nearly negligible.

Moreover, the degree of internal curing was observed as causing changes in the development of the autogenous shrinkage over time beyond the first few days. For the period beginning at one or two days, on the one hand, it was observed that the strain curves for UHPC with 0.3% SAP followed generally the trend of the corresponding curves for UHPC without SAP addition, thus showing approximately the same increase in autogenous shrinkage over time. On the other hand, UHPC with higher SAP additions (S0.4_1 and S1.0_1) showed only a very minor increase in

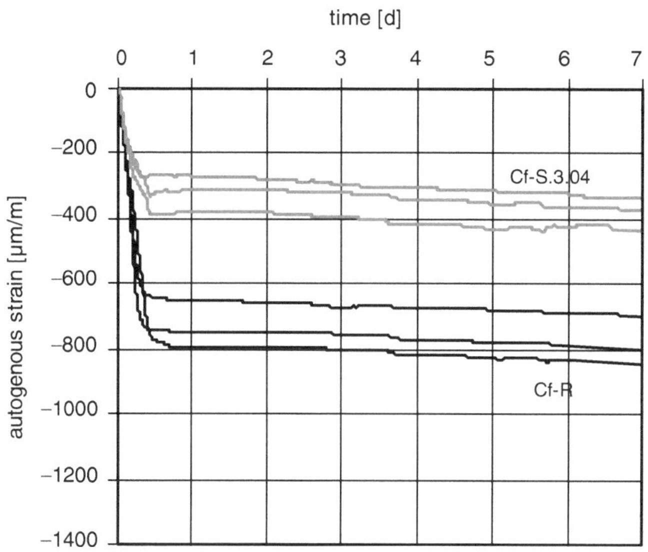

Fig. 7.18 Autogenous shrinkage of coarse grained UHPC measured starting after the final set: Cf-R = reference mixture with w/b= 0.21; Cf-S.3.04 = mixture with the addition of 0.3% SAP by mass of cement and extra water corresponding to $w_{IC}/c=0.04$ [47]

autogenous deformations for the same time interval. This indicates that the curing effect of the entrained water continued to work even at concrete ages higher than a mere first few days, as seems to be the case for UHPC with a considerably smaller amount of SAP and extra water; obviously in the later case the available amount of curing water was consumed already on the first day after final set.

Another investigation was performed on an UHPC composition containing coarser aggregates (basalt, maximum aggregate size of 8 mm) and 192 kg of steel fibres per m^3 of concrete [47]. The reference mixture contained 650 kg/m^3 of Portland cement and 177 kg/m^3 of silica fume. The water-to-binder ratio was 0.21. The cube compressive strength at the age of 28 days was 188 MPa. Figure 7.18 shows individual curves obtained from the measurements on the UHPC mixes with and without SAP and extra water for the first seven days after the final set. The 7-day strain values for autogenous shrinkage in the reference mixture Cf-R are approximately 35% lower in comparison to the corresponding values for the previously presented fine-grain reference mixture Ref (cf. Figure 7.17). This is likely due to the lower binder content of the coarser mix as well as the presence of stiffer coarse aggregates and steel fibres, which counteracted the shrinkage deformations of the matrix to some extent. The addition of SAP (0.3% by mass of cement) and extra water ($w_{IC}/c=0.04$) led to a reduction of autogenous shrinkage deformations to approximately the same level as that measured in the fine-grain mixture with the same percentage of SAP and IC water addition (cf. Figure 7.17, mix S0.3_2).

7.5 Effect of SAP on Drying Shrinkage

7.5.1 Cement paste

The work of Jensen and Hansen [28] is among the few studies on the effect of internal curing on both autogenous and drying shrinkage. All the cement pastes had a basic w/c of 0.3 and were exposed to drying (RH=50%, T=20ºC) at an age of approximately 4.5 months by taking the specimens out of their corrugated moulds. Figure 7.19 shows that the drying shrinkage increased for the water-entrained pastes, contrarily to autogenous shrinkage, which decreased drastically. However, the somewhat "negative" effect of water entrainment on drying shrinkage did not adversely affect the total deformations, which clearly decreased with increasing amounts of SAP and extra water in the mixtures.

7.5.2 Concrete

Moennig and Reinhardt [48] investigated the effect of internal curing on total shrinkage of concrete with a water-to-cement ratio of 0.36 (REF036). In the mixture with SAP, 0.7% by mass of cement of suspension-polymerised particles was added to the mixture and the total amount of water was increased to w/c=0.42 (i.e., 0.36+0.06) in order to provide internal curing water. Another mixture with internal curing contained pre-saturated LWA. A second reference composition was produced with w/c=0.42 (REF042), representing the total available water content. The concrete specimens had dimensions of 530 mm×100 mm×100 mm. The measurements

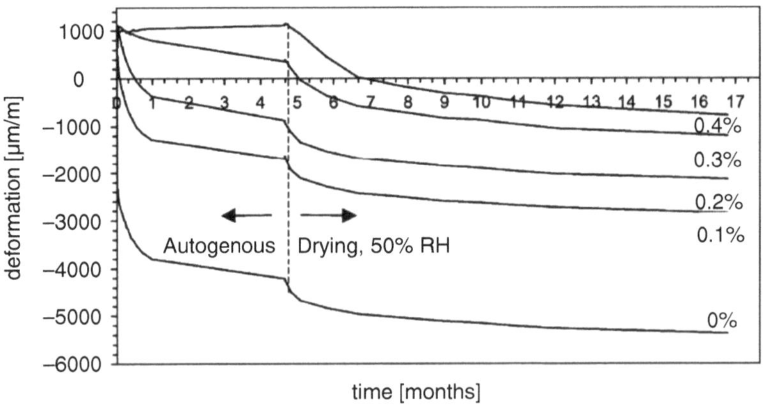

Fig. 7.19 Autogenous deformations measured from the setting time from drying shrinkage of cement pastes with different amounts of SAP and, consequently, different amounts of entrained water. The basic for all mixes was w/c = 0.3 [28]

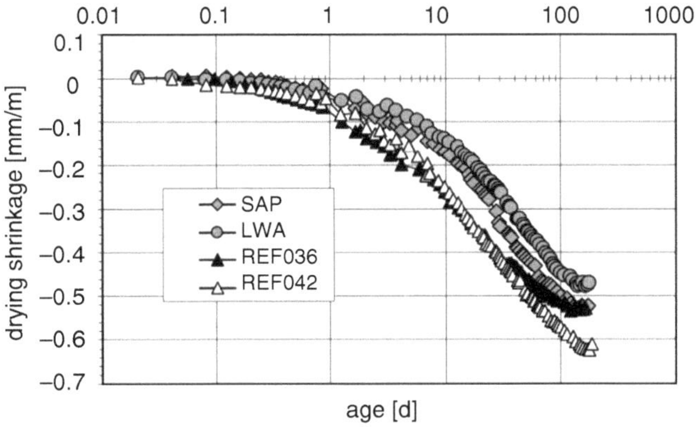

Fig. 7.20 Shrinkage deformations of four concrete mixtures. The points of curve are the individual average of two specimens and four measurements [48]

started immediately after demoulding, i.e., 24 hours after casting. At that time the specimens were exposed on all sides to the ambient atmosphere (temperature of 21°C and relative humidity of 45%). Figure 7.20 shows the results obtained, where one can observe that the influence of SAP resulted in decreased shrinkage values in comparison to the absence of SAP in both reference concretes. The authors explained the shrinkage-reducing effect of SAP on the shrinkage rate of concrete within the first week by the densification of the paste and the binding of the water by the polymers.

In addition to shrinkage measurements, the weight loss of the mixtures was recorded over time. After 180 days, the reference mixture REF042 and the mixture with SAP addition had approximately the same mass loss of 2.2%, while the reference mixture with the lower initial total water content (REF036) had a loss mass of only 1.5%.

Mechtcherine et al. [35] studied the effect of internal curing on total shrinkage of fine-grain UHPC, including drying shrinkage, by measuring the deformations of unsealed prisms with dimensions of 160 mm x 40 mm x 40 mm. The tests were executed at a temperature of 20°C and a relative humidity of 65%. The reference mixture, with a water-to-binder ratio of 0.19, is fully described in Section 7.4.3. The addition of 0.3% SAP by mass of cement (mixtures S0.3_1 and S0.3_2) was accompanied by the addition of extra water in the amounts of $w_{IC}/c=0.04$ and $w_{IC}/c=0.05$, respectively. The mixture S0.6_1 with 0.6% SAP had a content of internal curing water corresponding to $w_{IC}/c=0.08$. The test results for autogenous shrinkage obtained from the measurements on corrugated tubes and sealed prisms were presented in Figure 7.7a. Note that for the measurement of total shrinkage, the same prismatic specimen geometry was used as the one used for the corresponding autogenous shrinkage tests, cf. Figure 7.7b and 7.8b. Figure 7.21 shows the average curves of the total strains measured for the first 28 days after mixing concrete. An increase in total shrinkage was observed for the recorded time interval when SAP and additional curing water were used. More specifically, an increase in the amount

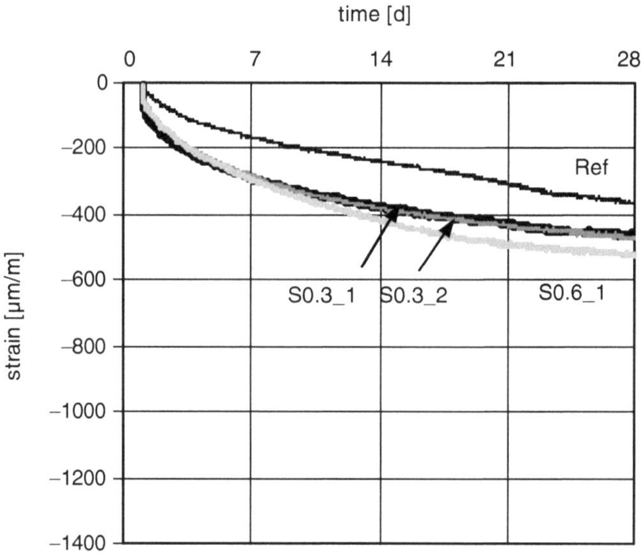

Fig. 7.21 Effect of the addition of SAP and IC water on total shrinkage of UHPC measured after demoulding the specimens at a concrete age of one day. Each curve is an average of two specimens [35]

of curing water resulted in an increase in total shrinkage. Taking into account the results for autogenous shrinkage (cf. Figure 7.7b) this increase can only be attributed to significantly higher drying shrinkage in the case of concretes with internal curing. This can be explained reasonably by the presence of a larger amount of free water present in such SAP-enriched concretes. The increase in total shrinkage due to internal curing is, however, much less pronounced than the reduction in autogenous shrinkage of concretes with SAP and extra water (cf. Figure 7.7a). It is also worth noting that the exposure of slender specimens to drying at the age of one day represents an extremely early beginning of desiccation and rather severe drying conditions.

In the continuation of the study presented above, the authors investigated the effect of age of exposure to desiccation [40, 46]. Again, prisms with the dimensions of 160 mm x 40 mm x 40 mm were used. Among the concretes examined were a control mix F-R, a mix F-S.4 containing 0.4% SAP but extra water, an internally cured concrete F-S.4.07 that contained 0.4% SAP and the amount of IC water corresponding to $w_{IC}/c=0.07$, and a mix F-R.07-1 which contained no SAP but extra water equal to w/c=0.07. In this mixture, less superplasticizer than in the corresponding mix F-S.4.07 was used in order to keep similar consistencies and to prevent bleeding. Figure 7.22 shows the average total strain curves obtained for these compositions which were exposed to drying at the concrete ages of 1, 7, 14, and 28 days, respectively.

The mixture F-S.4 containing SAP but no extra water showed a considerably lower total shrinkage in comparison to the reference mix F-R (Figure 7.22a).

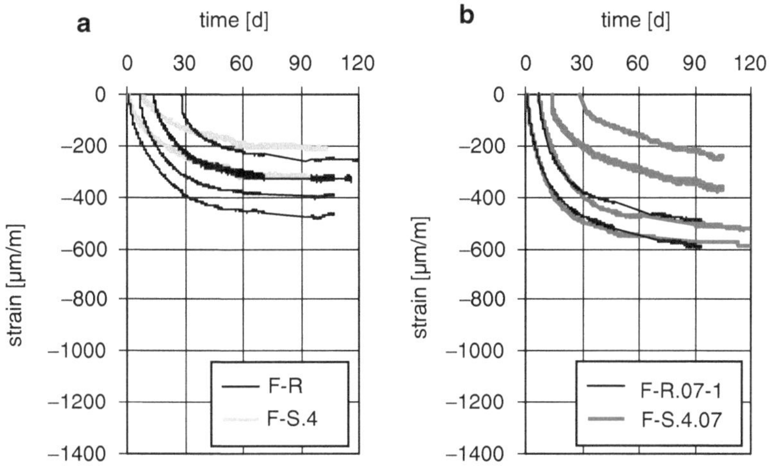

Fig. 7.22 Total shrinkage deformations for different exposure times to drying of: (**a**) reference UHPC mix F-R without SAP and concrete F-S.4 containing 0.4% SAP but no extra water, and (**b**) UHPC mix F-S.4.07 with 0.4% SAP and extra water corresponding to $w_{IC}/c=0.07$, as compared to the SAP-free mix containing an equivalent amount of extra water [40]

In contrast the total shrinkage deformations increased for the mix F-S.4.07 (cf. Figure 7.22b) containing SAP and extra water for internal curing, however, only for the exposure ages of 1 and 7 days. In the case of exposure to desiccation at later ages (14 or 28 days), nearly no difference in the total shrinkage deformations was observed in comparison to the mix F-R. It was pointed out by the authors that the decreasing difference in the deformations recorded between F-R and F-S.4.07 with increasing age of exposure manifests the depletion of internal curing water sources.

7.6 Development of Stresses due to Restraint

The test results presented in the foregoing sections showed free, non-restricted deformations of cement pastes and concretes made with and without SAP. The magnitudes of these deformations are expected to give an indication on the possible intensities of tensile stresses which would be found in concrete structures if the shrinkage strains were restrained. However, the real values of tensile stresses depend also on the development of concrete stiffness, creep deformations and on thermal dilation due to the heat of hydration. In order to estimate the actual tensile stresses due to restraint, two types of apparatus are commonly used: uniaxial-restrained shrinkage testing machine [49], and an instrumented ring [50].

In parallel to their investigations on free autogenous shrinkage, Igarashi and Watanabe [30] studied the development of stresses due to restrained shrinkage using a closed-loop, computer-controlled, uniaxial-restrained shrinkage (CLCCURS) test apparatus, cf. Figure 7.23 [49]. The mixture proportions and the test conditions are

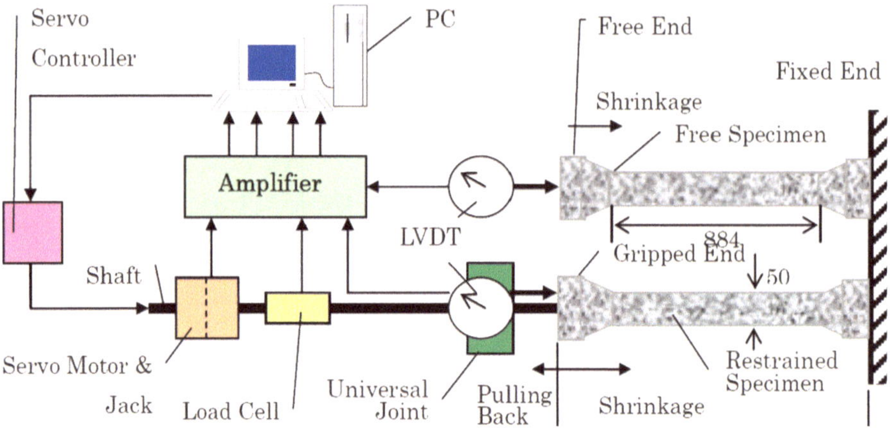

Fig. 7.23 Uniaxial restrained-shrinkage (CLCCURS) testing machine [49]

presented in the previous Sections 7.4.1 and 7.4.3. According to [30], the addition of SAP successively reduced stresses due to restraint. No tensile stress was generated in the concrete with 0.70% SAP, which was consistent with the complete elimination of free autogenous shrinkage in this mix. The tensile stresses measured for the reference mixtures were between 0.7 and 1.1 MPa at an age of 7 days. It should be mentioned that unexpectedly the stress in the reference concrete with the water-to-cement ratio of 0.3 was greater than the stress developed in the concrete with the w/c = 0.25. This finding requires further investigation. For the mix containing 0.35% SAP, the tensile stress at the age of 7 days reached 0.3 MPa.

Jensen and Hansen [28] evaluated the cracking susceptibility of mortars with and without SAP addition using an instrumented ring test. The setup typically consists of an inner steel ring that partially restrains the autogenous shrinkage of the annular test specimen cast around the ring. Due to the specimen shrinkage, this type of restraint gives rise to a uniform, radial pressure on the steel ring and induces tensile hoop stresses in the specimen. The stress build-up can be measured by strain gauges on the inside of the inner steel ring. Figure 7.24 shows the mortar test results. Significant tensile hoop stresses are induced in the mortar without SAP addition: After approximately 3 days of hardening, this mortar cracked as indicated by the sudden drop in stress. Internal curing based on SAP addition is seen to be effective in counteracting the stress build-up, which was counteracted partly in the specimen with 0.3% SAP addition and almost fully in the specimen with 0.6% SAP addition. The small stress, approximately 0.15 MPa, measured for the 0.6% SAP addition may have been induced by the slight temperature change during hydration. In any case, neither of the two mortars at 0.3% and 0.6% SAP addition cracked within the 3-week period of testing.

Mechtcherine et al. [46, 47] also used the instrumented ring test to assess the magnitude of the tensile stresses developed due to restrained autogenous deformations.

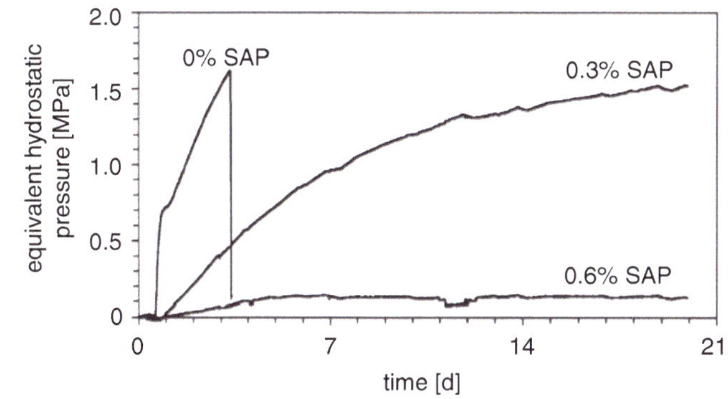

Fig. 7.24 Equivalent hydrostatic pressure developed in the inner ring by mortars with different amounts of SAP and, consequently, different amounts of entrained curing water. SAP additions are given by weight of cement, where 0.6% corresponds to an entrained w/c of 0.075. The basic w/c was 0.3 for all mixtures [28]

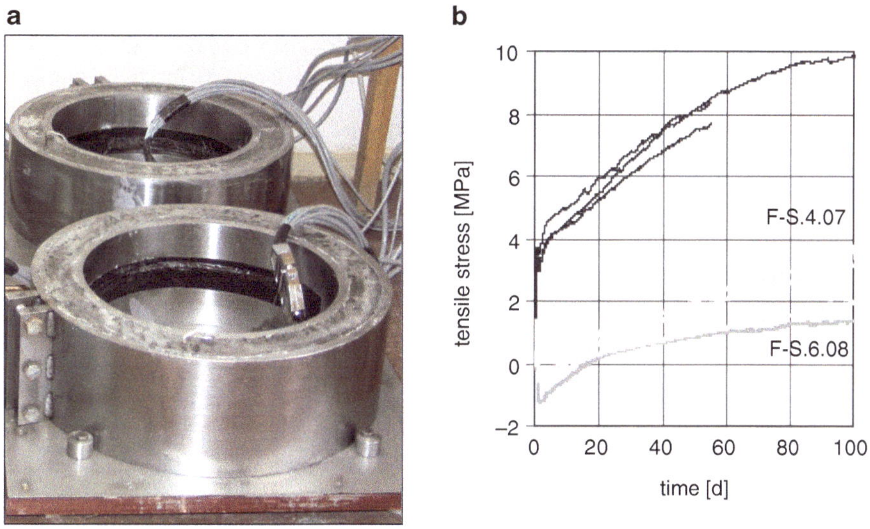

Fig. 7.25 (**a**) Setup for the instrumented ring test; (**b**) development of tensile stresses due to restraint in sealed concrete specimens made of fine-grained UHPC without SAP addition (F-R) and with SAP (F-S.4.07: SAP 0.4% of cement mass, $w_{IC}/c=0.07$; F-S.6.08: SAP 0.6% of cement mass, $w_{IC}/c=0.08$) in the first 100 days after casting. The graph includes data from [46, 47] supplemented by additional measurements from [52]

They estimated quantitatively the cracking tendency of fine-grain UHPC with and without SAP addition. A concrete annulus was cast around a steel ring of dimensions standardized by ASTM C1581-04 [50], cf. Figure 7.25a. The strains were measured continuously by four strain gauges glued to the inner side of the steel ring. These strain values were used for calculating the tensile stresses in concrete according to the

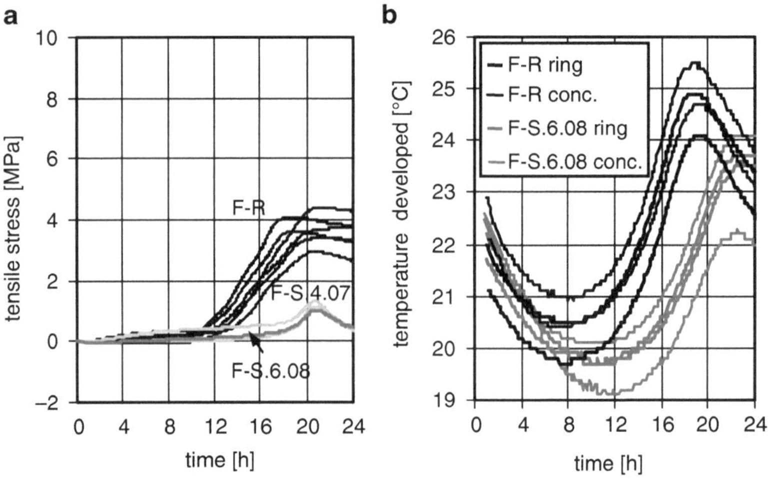

Fig. 7.26 Results of instrumented ring tests on sealed concrete specimens made of fine-grained UHPC without SAP addition (F-R) and with SAP (F-S.4.07: SAP 0.4 mass-% of cement, $w_{IC}/c=0.07$; F-S.6.08: SAP 0.6 mass-% of cement $w_{IC}/c=0.08$) in the first 24 hours after casting: (**a**) development of tensile stresses due to restraint in numerous tests performed [46, 47]; (**b**) typical temperature changes measured in concrete and at the inner surface of the inner steel ring due to boundary conditions and hydration heat [52]

equations proposed in [51]. Figure 7.25b shows the results obtained from ring tests for the reference concrete F-R and SAP-enriched mixtures F-S.4.07 (SAP 0.4% of cement mass, $w_{IC}/c=0.07$) and F-S.6.08 (SAP 0.6% of cement mass, $w_{IC}/c=0.08$) over the first 100 days after mixing. The mitigation of autogenous shrinkage using internal curing caused a dramatic reduction of the stresses due to restraint. For example, at a concrete age of 50 days, the stresses induced in the specimens made with the mix F-S.4.07 containing SAP were approximately 2 MPa and so 4 times smaller than the corresponding stresses in the reference concrete at the same age. For a higher SAP content and a higher content of additional water used in mixture F-S.6.08, the developed stresses were even lower. This clearly demonstrates the considerable reduction of the cracking potential of UHPC as a result of internal curing.

As can be seen in Figure 7.25b, the development of stresses is very unsteady at early concrete ages, which likely results from temperature changes due to the heat of hydration; see also Figure 7.26a showing the tensile stress levels in the first 24 hours after casting. In order to estimate the possible effects of hydration heat on stress development, temperature was measured on the inner steel ring and in concrete. Figure 7.26b shows the development of temperature in the experiments on the fine-grain UHPC with and without internal curing produced and cast on the same day. Note that no calibration of the measuring sensors with regard to absolute temperature was carried out because it was not necessary, i.e., only the relative temperature changes need be considered. The minimum temperatures recorded with each sensor (the lowest point of each curve) roughly corresponded to the room temperature in the lab (approximately 20°C).

Both concretes, which warmed up during the intensive mixing process, showed a gradual decrease in temperature during the first 7-8 hours (mixture without SAP) and approximately in the first 10-11 hours (mixture with SAP) after casting, which was due to the cooling of the mixes down to the temperature of the steel rings and the ambient atmosphere in the lab. Subsequently, the temperature rose again due to the heat of hydration and reached its maximum at a concrete age of approximately 18-19 hours for the reference mixture and at about 22-23 hours for the concrete containing SAP and extra water. It appears as well that the rise of temperature is less pronounced when internal curing is used.

The extent of the thermal stresses which can be developed in a concrete ring test depends mainly on the difference in the coefficients of thermal expansion (CTE) of steel and concrete. Given that the CTE of the steel is higher than that of the concrete, any increase in temperature in the system would induce tensile stresses in concrete, and cooling would cause the opposite effect. In the experiments cited above, the ring steel had a CTE of 12×10^{-6} 1/K according to the steel producer [52]. The CTE of UHPC with and without internal curing depends – as for any other concrete – on the concrete age and the type of aggregates. The authors of [46, 47] did not measure the CTE of the UHPC, but this was done in another study conducted on very nearly the same UHPC compositions, one of which also contained SAP, indeed without extra water [53]. The measured CTE values for the very young UHPC were between 11.0 and $11.5 * 10^{-6}$ 1/K, and hence slightly below that of the steel used in fabrication of the rings. Even if the difference in the CTE was not pronounced, it seems to have been large enough to induce thermal stresses due to the hydration heat observed in the experiments.

In order to minimise the effects of temperature, Eppers et al. [54] modified their test setup by reducing the rectangular cross-section of the concrete ring to 24 mm x 25 mm, and the geometry of the steel ring was adapted accordingly. The maximum temperature rise did not exceed 1.5K in their test. The authors worked with the same UHPC reference composition (indicated with 1A in Figure 7.27) as in [46, 47] with only very slight changes in mix proportions. Furthermore, another type of SAP at a dosage of 0.3% by mass of cement was used in the SAP-enriched mixture (1A SAP), and no extra water was added. Additionally, a UHPC mixture with a shrinkage-reducing admixture (mixture 1A SRA) was investigated. Figure 7.27 shows the development of tensile stresses in the concrete specimens and their estimated cracking propensity, which was defined as ratio of the tensile stress and the measured or converted splitting tensile strength. According to these results, the addition of SAP leads to a considerable reduction of stresses developed due to restrained autogenous shrinkage, and the cracking propensity is diminished as well. It is worth noting that the use of an SRA agent also provided good results with regard to the mitigation of autogenous shrinkage and its consequences.

In their study of high-strength mortars, Schlitter et al. [55] dealt with both the effects of internal curing by using SAP and of temperature change on the stress

[3] Radocea A (1992) Study on the mechanism of plastic shrinkage of cement-based materials. PhD thesis, Chalmers University of Technology, Sweden
[4] Radocea A (1992) A new method for studying bleeding of cement paste. Cem Concr Res 22(5):855–868
[5] Kratz M, Dudziak L, Mechtcherine V (2007) Plastic shrinkage of concretes with low and high w/c ratios. Internal technical report, TU Dresden
[6] Dudziak L, Mechtcherine V (2010) Enhancing early-age resistance to cracking in high-strength cement-based materials by means of internal curing using Super Absorbent Polymers. In: Brameshuber W (ed) Proceedings of international conference on Material Science, 6-10 September (Aachen University, Aachen, Germany), RILEM Proc. PRO 77, Vol III (Additions Improving Properties of Concrete-AdIPoC) RILEM Publications S.A.R.L., Bagneux, France,129–139
[7] Dudziak L (2011) Mitigating autogenous shrinkage of high-performance concrete by using Super Absorbent Polymers. PhD thesis, TU Dresden, Germany (in preparation)
[8] Le Chatelier H (1900) Sur les changements de volume qui accompagnent le durcissement des ciments. Bull Soc Encour Ind Natl V (5th series), 54–57
[9] Jensen OM, Hansen PF (2001) Autogenous deformation and RH-change in perspective. Cem Concr Res 31(12):1859–1865
[10] Bouasker M, Mounanga P, Turcry P., Loukili A, Khelidj A (2008) Chemical shrinkage of cement pastes and mortars at very early age: Effect of limestone filler and granular inclusions. Cem Concr Compos 30(1):13–22
[11] Nawa T, Horita T (2004) Autogenous shrinkage of high-performance concrete. In: Proceedings of the international workshop on Microstructure and Durability to Predict Service Life of Concrete Structures, February 2004 (Sapporo, Japan), accessed online: http://www.hucc.hokudai.ac.jp/~m16120/coe/workshop/nawa4.pdf
[12] Powers TC, Brownyard TL (1948) Studies of the physical properties of hardened Portland cement paste (9 parts). J Amer Concr Inst 43 (Oct. 1946 to April 1947), Bulletin 22, Research Laboratories of the Portland Cement Association, Chicago
[13] Jensen OM (1993) Autogenous deformation and RH-change - self-desiccation and self-desiccation shrinkage, PhD thesis, Technical University of Denmark, Denmark (in Danish).
[14] Hansen TC (1986) Physical structure of hardened cement paste. A classical approach. Mater Struct 19(6):423–436
[15] Beltzung F, Wittmann FH (2001) Early chemical shrinkage due to dissolution and hydration of cement. Mater Struct 34(5):279–283
[16] Lura P (2003) Autogenous deformation and internal curing of concrete. PhD thesis, TU Delft, The Netherlands
[17] Justnes H, Sellevold EJ, Reyniers B, Van Loo D, Van Gemert A, Verboven F et al. (2000) Chemical shrinkage of cement pastes with plasticizing admixtures. Nordic Concr Res 24:39–54
[18] Japan Concrete Institute Report (1998) Technical committee on autogenous shrinkage of concrete. In: Proceedings of the international workshop on autogenous shrinkage of concrete autoshrink '98, Hiroshima (Japon). Londres: E&FN Spon, 3–62.
[19] Esteves LP (2009) Internal curing in cement-based materials. PhD thesis, Universidade de Aveiro
[20] Lyman CG (1934) Growth and Movement in Portland Cement Concrete, Oxford University Press, London
[21] DIN 1045-1 (2008) Tragwerke aus Beton, Stahlbeton und Spannbeton – Teil 1: Bemessung und Konstruktion. Beuth Verlag GmbH, Berlin, Germany
[22] DIN EN 1992-1-1 Berichtigung 1, Januar 2010. Eurocode 2: Bemessung und Konstruktion von Stahlbeton- und Spannbetontragwerken - Teil 1-1: Allgemeine Bemessungsregeln und Regeln für den Hochbau; Deutsche Fassung EN 1992-1-1:2004, Berichtigung zu DIN EN 1992-1-1:2005-10; Deutsche Fassung EN 1992-1-1:2004/AC:2008, Beuth Verlag GmbH, Berlin, Germany
[23] JSCE (2002) Standard specifications for design and construction of concrete structures

[24] RILEM Report 41 (2007) State of the Art Report of RILEM Technical Committee Internal Curing of Concrete. Kovler K, Jensen OM (eds), RILEM Publications S.A.R.L., Bagneux, France
[25] Weber S, Reinhardt HW (1997) A New Generation of High Performance Concrete: Concrete with Autogenous Curing. Adv Cem Based Mater 6 (2):59–68
[26] Bentur A, Igarashi S, Kovler K (2001) Prevention of Autogenous Shrinkage in High-Strength Concrete by Internal Curing Using Wet Lightweight Aggregates. Cem Concr Res 31(11):1587–1591
[27] Jensen OM, Hansen PF (2001) Water-Entrained Cement-Based Materials: I. Principle and Theoretical Background. Cem Concr Res 31(4):647–654
[28] Jensen OM, Hansen PF (2002) Water-Entrained Cement-Based Materials: II. Experimental observations. Cem Concr Res 32(6):973–978
[29] Mechtcherine V, Dudziak L, Schulze J, Stähr H (2006) Internal curing by Super Absorbent Polymers – Effects on material properties of self-compacting fibre-reinforced high performance concrete. In: Jensen OM, Lura P, Kovler K (eds) Proceedings of international RILEM conference on Volume Changes of Hardening Concrete: Testing and Mitigation, 20-23 August 2006 (Technical University of Denmark, Lyngby, Denmark), 87–96
[30] Igarashi S, Watanabe A (2006) Experimental study on prevention of autogenous deformation by internal curing using super-absorbent polymer particles. In: Jensen OM, Lura P, Kovler K (eds) Proceedings of international RILEM conference on Volume Changes of Hardening Concrete: Testing and Mitigation, 20-23 August 2006 (Technical University of Denmark, Lyngby, Denmark), 77–86
[31] NF P 18-427 (1996) Determination of the dimensional variations between two opposite faces of hardened test specimen concrete. French standards, AFNOR, Paris, France
[32] Jensen OM, Hansen PF (1995) A dilatometer for measuring autogenous deformation in hardening Portland cement paste. Mater Struct 28(181):406–409
[33] ASTM International, Committee (2010) Standard ASTM C 1698-09. In: Annual Book of ASTM Standard, C-09 Volume 04.02 – Concrete and Aggregates, Section 04- Construction, West Conshohocken, USA
[34] DIN 52450 (1985) Prüfung anorganischer nichtmetallischer Baustoffe; Bestimung des Schwindens und Quellens an kleinen Probekörpern, Beuth Verlag GmbH, Berlin, Germany
[35] Mechtcherine V, Dudziak L, Hempel S (2009) Mitigating early age shrinkage of concrete by using Super Absorbent Polymers (SAP). In: Tanabe T et al. (eds) Proceedings of the 8th international conference on Creep, Shrinkage and Durability Mechanics of Concrete and Concrete Structures – CONCREEP-8, 30 September-2 October 2008 (Ise-shima, Japan), 847–853
[36] DIN EN 480-2 (2006) Zusatzmittel für Beton, Mörtel und Einpressmörtel - Prüfverfahren - Teil 2: Bestimmung der Erstarrungszeit; Deutsche Fassung EN 480-2:2006, Beuth Verlag GmbH
[37] Sant G, Lura P, Weiss J (2006) A discussion of analysis approaches for determining 'time-zero' from chemical shrinkage and autogenous strain measurements in cement paste. In: Jensen OM, Lura P, Kovler K (eds) Proceedings of international RILEM conference on Volume Changes of Hardening Concrete: Testing and Mitigation, 20-23 August 2006 (Technical University of Denmark, Lyngby, Denmark), 375–383
[38] Sant G, Dehadrai M, Lura P, Bentz D, Ferraris CF, Bullard JW, Weiss J (2009) Detecting the fluid-to-solid transition in cement pastes: Assessment techniques. Concr Int 31(06):54–58?
[39] Amziane S (2006) Setting time determination of cementitious material based on measurements of the pore water pressure variations. Cem Concr Res 36(2):295–304
[40] Dudziak L, Mechtcherine V (2010) Deliberations on kinetics of internal curing water migration and consumption based on experimental studies on SAP-enriched UHPC. Int RILEM Conference on Use of Superabsorbent Polymers and Other New Additives in Concrete, 15-18 August 2010, Technical University of Denmark, Lyngby, RILEM Proceedings PRO 74, 33–43
[41] Ribeiro AB (2009) Personal communication, March 22.

[42] ASTM C1608 - 07 Standard Test Method for Chemical Shrinkage of Hydraulic Cement Paste
[43] Lura P, Durand F, Jensen OM (2006) Autogenous strain of cement pastes with superabsorbent polymers. In: Jensen OM, Lura P, Kovler K (eds) Proceedings of international RILEM conference on Volume Changes of Hardening Concrete: Testing and Mitigation, 20-23 August 2006 (Technical University of Denmark, Lyngby, Denmark), 57–66
[44] Geiker M, Bentz DP, Jensen OM (2004) Mitigating autogenous shrinkage by internal curing. High Performance Structural Lightweight Concrete, ACI SP-218, Ries JP, Holm TA (eds), 143–154
[45] Dudziak L, Mechtcherine V (2008) Mitigation of volume changes of Ultra-High Performance Concrete (UHPC) by using Super Absorbent Polymers. In: Fehling E et al (eds.) Proc of the 2nd international symposium on Ultra High Performance Concrete, 5-7 March 2008 (Kassel, Germany), Kassel University Press GmbH, 425–432
[46] Mechtcherine V, Dudziak L, Hempel S (2009) Reducing cracking potential of Ultra High Performance Concrete by internal curing using Super Absorbent Polymers. In: Kovler K (eds) Proceedings of 2nd international RILEM workshop on Concrete Durability and Service Life Planning, ConLife'09, 7-9 September 2009 (Haifa, Israel), 31–38
[47] Dudziak L, Mechtcherine V (2010): Reducing the cracking potential of Ultra-High Performance Concrete by using Super Absorbent Polymers (SAP). In: Van Zijl GPAG, Boshoff WP (eds) Proceedings of the international conference on Advanced Concrete Materials, 17-19 November 2009 (Stellenbosch University, Stellenbosch, South Africa), 11–19
[48] Reinhardt HW, Mönnig S (2006) Results of comparative study of the shrinkage behaviour of concretes with different internal water sources. In: Jensen OM, Lura P, Kovler K (eds) Proceedings of international RILEM conference on Volume Changes of Hardening Concrete: Testing and Mitigation, 20-23 August 2006 (Technical University of Denmark, Lyngby, Denmark), 67–76
[49] Kovler K (1994) Testing system for determining the mechanical behaviour of early age concrete under restrained and free uniaxial shrinkage. Mater Struct 27(6):324–330
[50] ASTM Standard C1581-04 Standard Test Method for Determining Age at Cracking and Induced Tensile Stress Characteristics of Mortar and Concrete under Restrained Shrinkage
[51] Hossain AB, Weiss WJ (2004) Assessing residual stress development and stress relaxation in restrained concrete ring specimens. Cem Concr Compos 26(5):531–540
[52] Dudziak L: Personal communication, May 2010
[53] Eppers S, Mueller C (2009) On the examination of the autogenous shrinkage cracking propensity by means of the restrained ring test with particular consideration of the temperature influences. VDZ Concrete Technology Reports 2007-2009, 57–70
[54] Eppers S, Mechtcherine V, Mueller Ch (2010) Assessing the autogenous shrinkage cracking propensity of concrete- Methodological aspects. In: Jensen OM, Hasholt MT, Laustsen S (eds) Proceedings of international RILEM conference on Use of Superabsorbent Polymers and Other New Additives in Concrete, 15-18 August 2010 (Technical University of Denmark, Lyngby, Denmark), 45–56
[55] Schlitter JL, Barrett T, Weiss J (2010) Restrained shrinkage behaviour and thermal effects in mortars containing super absorbent polymers (SAP). In: Jensen OM, Hasholt MT, Laustsen S (eds) Proceedings of international RILEM conference on Use of Superabsorbent Polymers and Other New Additives in Concrete, 15-18 August 2010 (Technical University of Denmark, Lyngby, Denmark), 233–242

Chapter 8
Effect of Superabsorbent Polymers on the Mechanical Properties of Concrete

Konstantin Kovler

Abstract The present chapter deals with the effect of internal curing by means of SAP on mechanical properties of hardened concrete, such as compressive and tensile strength under compressive and different kind of tensile loading (uniaxial, bending and splitting), and elastic properties.

8.1 Introduction

The present chapter deals with the effect of internal curing by means of SAP on mechanical properties of hardened concrete, such as compressive and tensile strength under compressive and different kind of tensile loading (uniaxial, bending and splitting), and elastic properties. Unfortunately, the information on other mechanical properties of concrete containing SAP, such as creep and fracture mechanics characteristics, is still not available.

The focus in this chapter is made on the effect of SAP introduction on concrete strength (mainly, compressive). At the same time, a few researchers reported data on tensile strength and elasticity modulus, which are reviewed hereafter as well.

It has to be emphasized that most of the SAP uses in concrete construction are based on its positive effect as internal curing (IC) agent in high-strength concrete and aimed to mitigate self-desiccation and autogenous shrinkage. An accompanying effect, some moderate strength reduction would be expected in early ages, because of higher porosity and moisture of the specimens (if they are tested after sealed curing, it means not under one of the two extreme moisture conditions: complete water-saturated or oven-dried). At the same time, the specimens tested at

K. Kovler (✉)
National Building Research Institute - Faculty of Civil and Environmental Engineering, Technion - Israel Institute of Technology, Haifa, Israel
e-mail: cvrkost@techunix.technion.ac.il

V. Mechtcherine and H.-W. Reinhardt (eds.), *Application of Superabsorbent Polymers (SAP) in Concrete Construction*, RILEM State of the Art Reports 2,
DOI 10.1007/978-94-007-2733-5_8,

mature ages sometimes indicate higher strength and elasticity modulus than those made of reference mixtures (without SAP). This improvement is probably due to enhanced cement hydration, which can compensate strength reduction caused by slightly higher porosity. The results of the studies, where SAP is used as IC agent, are reviewed in following sections. This information is grouped by mechanical properties of hardened concrete.

Another possible application is introducing SAP as water retaining agent. As a result of a fast water absorption in the first minutes of mixing after contact with water, an effective water/cement ratio in the cement matrix decreases, and concrete strength of the hardened concrete increases. In general, tensile strength and elasticity modulus are expected to increase as well, however it does not always happen. These data are assembled and reviewed in the end of the chapter.

The studied compositions are described, whenever is possible, by the basic water to cement ratio (w/c), entrained (for the purposes of IC) water to cement ratio (w_e/c) and SAP content (% by cement mass).

8.2 Compressive Strength

Most of the publications report about reduction of compressive strength for concrete containing SAP, in comparison with reference concrete, especially at early ages. Few publications demonstrate almost the same or higher strength at later ages. In general, the resulting strength of concrete with SAP depends on curing conditions, age and material composition. The present section reviews a number of key publications, where the effect of SAP has been studied systematically.

Compressive strength measured on 45x90-mm mortar cylinders after 1 day of sealed curing followed by 27 days of water curing at 20°C, is reported by Jensen and Hansen in the work [1]. The average compressive strength was 134 MPa for a reference mortar (0% SAP) and 109 MPa for a water-entrained mortar (w/c=0.3, w_e/c=0.05, 0.6% SAP), i.e. the strength was reduced by 19% due to SAP addition.

Higher or equal later-age compressive strength compared to normal mortar was obtained by Bentz et al. in mortar with SAP, while extra water/cement ratio needed for internal curing was 0.046 [2].

Mönnig [3] concluded that compressive strength development of mortars prepared at different w/c using different types of SAP did not show any significant difference between mixtures with or without polymers. However, this statement was illustrated by the tests made at w/c=0.55 only, while the primary purpose was to study freezing-thawing resistance of normal-strength mortars. This ratio seems to be rather high and not typical for traditional SAP applications in high-strength cementitious materials.

Craeye and De Schutter [4] observed strength reduction of concrete made at w/c=0.32 contained silica fume and SAP (at different SAP contents) (Table 8.1).

As can be seen from this table, the strength reduction is more pronounced for higher SAP dosages and contents of entrained water w_e.

Table 8.1 Compressive strength of concrete with SAP at 28 days [4]

SAP, %	w_e/c	Compressive strength, MPa	Strength reduction, %
0.049	0.06	91.3	16
0.065	0.08	79.8	27
0.081	0.10	75.0	31

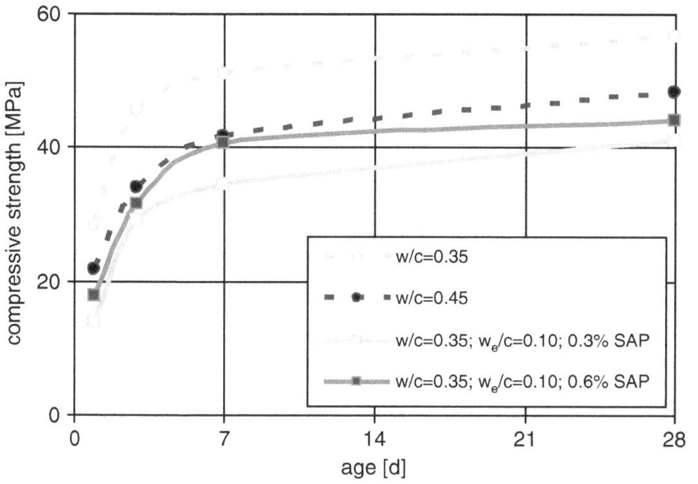

Fig. 8.1 Development of compressive strength in time of concrete with SAP in comparison with that of reference concrete mixes made at w/c=0.35 and 0.45, adapted from [6]

In another work conducted in the same laboratory [5] another type of SAP was used, but the significant reduction of 28 day compressive strength was obtained as well. The significant strength reduction found in these two works is not common for SAP-containing concrete. It seems that the authors overestimated water absorption of SAP, which was chosen 130 g/g [4] and 45 g/g [5], based on the absorption in water after 5 minutes. Adding too much mixing water into the concrete should inevitably reduce strength. To remind the readers, observations of SAP swelling in cement pastes indicate that the total absorption is about half the absorption for synthetic pore fluid, which in turn is several times less than the absorption of SAP in distilled water [1].

Hoa Lam and Hooton [6] studied the influence of different materials, which can be used for shrinkage mitigation, such as SAP, lightweight aggregate (LWA) sand and shrinkage-reducing admixture (SRA). Concrete mixes were made at w/c=0.35, while an additional concrete mixture with w/c=0.45 was included in the experimental to benchmark the changes in the concrete mixtures with SAP. It was assumed that to compensate an initial absorption of dry particles of SAP in fresh concrete mix, additional water should be added into the mixes. However, the amount of this water was the same (w_e/c=0.10) for two different contents of SAP, 0.3% and 0.6% by the cement mass, which was different from the common practice, when the amount of entrained water is directly proportional to the SAP concentration. The results of the strength measurements are given in Fig. 8.1. It can be seen that the introduction of

Table 8.4 Compressive strength of concrete cubes and prisms made with and without SAP and cured at different conditions [10]

	Compressive strength of cubes, MPa		Compressive strength of beam halves, MPa	
	20°C/28d	90°C/2d	20°C/28d	90°C/2d
Reference concrete	132	160	172	175
Concrete with SAP	129	144	150	169

pastes, the voids left by the largest SAP particles, about 0.5 mm across, are the largest defects. In the mortars, however, the largest defects are paste-aggregate interfaces, as well as weak or pre-cracked aggregates. These defects are usually of the size of aggregates (up to 2 mm in this work), i.e. significantly larger than the SAP particles. This may explain why no influence of the SAP on the compressive strength of the mortars could be detected. It may also explain the lower strength of the mortars compared to the cement pastes. The authors concluded that since aggregate or interfacial transition zones, and not cement paste, are the limiting factor for mortar strength, the influence of SAP on strength of mortar and concrete should be negligible.

Mechtcherine et al. [10] measured compressive strength of reference concrete made at w/c=0.25 and concrete with 0.4% of SAP (w/c=0.29, including internal curing water) on both 150-mm cubes. Two different curing conditions were applied for the specimens: storing unsealed in the standard laboratory climate and tested at the age of 28 days, and alternatively, storing sealed and treatment in an oven at a temperature of 90°C for 2 days. Table 4 gives the average values for the standard curing. The use of SAP resulted in some decrease in compressive strength. The same trends were obtained for compressive strength measured on halves of small beams with 40x40 mm cross-section (cf. Table 8.4), uniaxial tensile and flexural strength, except for tensile strength of oven-stored specimens.

Esteves et al. [11] tested mortar mixes at different w/c (from 0.25 to 0.35) under different curing conditions ranged from 30% to 100% of relative humidity at 20°C. They observed 20% reduction in compressive strength of SAP-containing mortars, when water curing was applied. At the same time, the results of testing mortars cured in adverse curing conditions were more favorable for SAP mixtures, which maintained their compressive strength. The authors concluded that mortars internally cured by means of SAP "seem to be insensitive to external relative humidity change in terms of mechanical resistance".

Mechtcherine et al. [12] reported about noticeable decrease in compressive strength for ultra high performance SAP-containing concretes (w/c=0.22) at early ages (1-7 days). However, the measurements at 28 days exhibited only a minor decrease in strength in cases when the addition of SAP and extra water were relatively small (0.3% SAP). A higher dosage of SAP (0.6%) and internal curing water led to a considerable decrease in strength.

Dudziak and Mechtcherine [13] observed a reduction of compressive strength of ultra high performance concrete (UHPC), but found "no pronounced negative effects

Table 8.1 Compressive strength of concrete with SAP at 28 days [4]

SAP, %	w_e/c	Compressive strength, MPa	Strength reduction, %
0.049	0.06	91.3	16
0.065	0.08	79.8	27
0.081	0.10	75.0	31

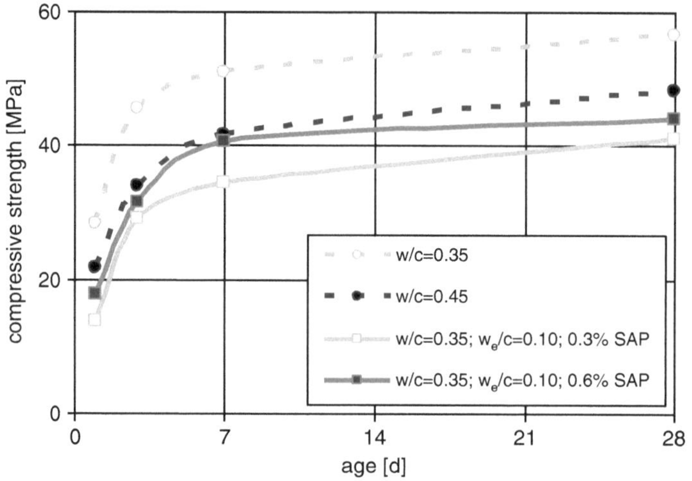

Fig. 8.1 Development of compressive strength in time of concrete with SAP in comparison with that of reference concrete mixes made at w/c=0.35 and 0.45, adapted from [6]

In another work conducted in the same laboratory [5] another type of SAP was used, but the significant reduction of 28 day compressive strength was obtained as well. The significant strength reduction found in these two works is not common for SAP-containing concrete. It seems that the authors overestimated water absorption of SAP, which was chosen 130 g/g [4] and 45 g/g [5], based on the absorption in water after 5 minutes. Adding too much mixing water into the concrete should inevitably reduce strength. To remind the readers, observations of SAP swelling in cement pastes indicate that the total absorption is about half the absorption for synthetic pore fluid, which in turn is several times less than the absorption of SAP in distilled water [1].

Hoa Lam and Hooton [6] studied the influence of different materials, which can be used for shrinkage mitigation, such as SAP, lightweight aggregate (LWA) sand and shrinkage-reducing admixture (SRA). Concrete mixes were made at w/c=0.35, while an additional concrete mixture with w/c=0.45 was included in the experimental to benchmark the changes in the concrete mixtures with SAP. It was assumed that to compensate an initial absorption of dry particles of SAP in fresh concrete mix, additional water should be added into the mixes. However, the amount of this water was the same (w_e/c=0.10) for two different contents of SAP, 0.3% and 0.6% by the cement mass, which was different from the common practice, when the amount of entrained water is directly proportional to the SAP concentration. The results of the strength measurements are given in Fig. 8.1. It can be seen that the introduction of

Table 8.2 Compressive strength of concrete with SAP at 7 and 28 days [7]

SAP, %	w_e/c	Compressive strength, MPa	
		7 days	28 days
0	0	86.3	97.4
0.35	0.045	78.0	94.2
0.70	0.09	67.0	76.1

Table 8.3 Compressive strength of concrete with SAP at 2, 7 and 28 days [8]

SAP, %	w_e/c	Compressive strength, MPa		
		2 days	7 days	28 days
0	0	82	97	107
0.3	0.02	70	84	99
0.6	0.04	65	76	93

SAP reduced compressive strength. The highest compressive strength was obtained for reference concrete made at w/c=0.35.The lower strength of the system with 0.3% SAP could be explained by too high content of entrained water, twice as much as that introduced per unit mass of super-absorbent polymer in the case of 0.6% SAP. The strength of the mix with 0.6% SAP was approximately equal to the strength of the reference mix made at w/c=0.45. It means that these mixtures had a similar "equivalent" porosity and microstructure, which are characterized by the combination of capillary porosity and hydration degree. Although the degree of hydration was not determined in this study, but we can assume that the negative influence of additional voids induced by the addition of SAP particles was compensated by the improved degree of hydration in the mix containing 0.6% SAP.

Igarashi and Watanabe [7] studied compressive strength of cement pastes and concrete made with and without SAP. The SAP used was powder of acrylamide/acrylic acid copolymer in the form of mono-sized spherical particles having diameter about 200 μm in the dry state. Its absorption capacity was about 13 times its dry weight (dry density=1.25 g/cm^3). It was clearly demonstrated that the addition of SAP resulted in the decrease in compressive strength of both paste and concrete. The lower the w/c, the more sensitive were the materials to the amount of SAP. The basic w/c ratio in concrete specimens was 0.25. Table 8.2 shows that the addition of SAP, especially in the larger amount of 0.70%, decreased the compressive strength of concrete as well.

Piérard et al. [8] measured compressive strength of concrete cubes cured at temperature of 20±2°C and minimum air relative humidity of 95% at ages of 2, 7 and 28 days. Concrete mixtures had basic w/c of 0.35 and contained silica fume. The average results are provided in Table 8.3.

As can be seen from Table 8.3, the early strength development (2-7 days) is somewhat slowed down with SAP, but the reduction in strength seems to decrease at later ages. After 28 days, the reductions in compressive strength are 7% and 13% for concrete with SAP content of 0.3% and 0.6% respectively. The authors assume that these losses may be attributed to the increased porosity of the concrete due to the hollow voids introduced by SAP particles. However, they speculate that the

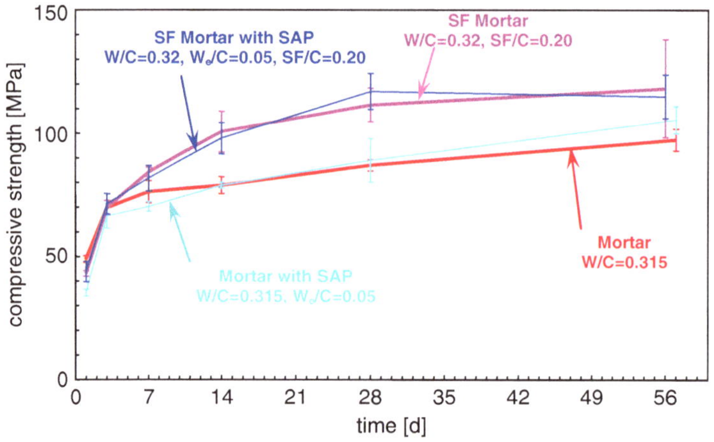

Fig. 8.2 Compressive strength (average strength values and standard deviation) of mortars with and without superabsorbent polymers [9]

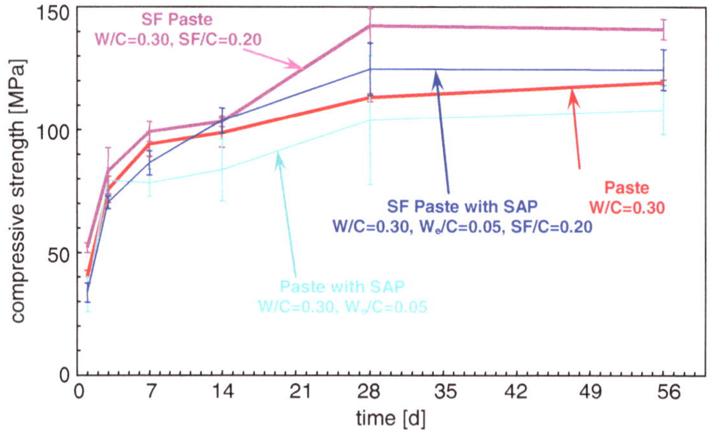

Fig. 8.3 Compressive strength (average strength values and standard deviation) of pastes made with and without superabsorbent polymers [9]

long-term strength can be similar or even higher, as the hydration proceeds at a higher rate since a higher internal RH is maintained in the specimens.

Lura et al. [9] measured strength on cylinders with diameter 24 mm and height of 50 mm, which were cured in sealed conditions at 20°C. Pastes and mortars, both with and without silica fume additive, were tested. It was found that internal curing by means of SAP had almost no influence on the compressive strength of high performance mortars (Fig. 8.2), while the compressive strength of cement pastes was reduced by 20% at early ages and by 10% at later ages (Fig. 8.3). The different effect of SAP introduction in cement pastes and mortars was explained by different largest defect size present in the solid matrix of these materials. For example, in cement

Table 8.4 Compressive strength of concrete cubes and prisms made with and without SAP and cured at different conditions [10]

	Compressive strength of cubes, MPa		Compressive strength of beam halves, MPa	
	20°C/28d	90°C/2d	20°C/28d	90°C/2d
Reference concrete	132	160	172	175
Concrete with SAP	129	144	150	169

pastes, the voids left by the largest SAP particles, about 0.5 mm across, are the largest defects. In the mortars, however, the largest defects are paste-aggregate interfaces, as well as weak or pre-cracked aggregates. These defects are usually of the size of aggregates (up to 2 mm in this work), i.e. significantly larger than the SAP particles. This may explain why no influence of the SAP on the compressive strength of the mortars could be detected. It may also explain the lower strength of the mortars compared to the cement pastes. The authors concluded that since aggregate or interfacial transition zones, and not cement paste, are the limiting factor for mortar strength, the influence of SAP on strength of mortar and concrete should be negligible.

Mechtcherine et al. [10] measured compressive strength of reference concrete made at w/c=0.25 and concrete with 0.4% of SAP (w/c=0.29, including internal curing water) on both 150-mm cubes. Two different curing conditions were applied for the specimens: storing unsealed in the standard laboratory climate and tested at the age of 28 days, and alternatively, storing sealed and treatment in an oven at a temperature of 90°C for 2 days. Table 4 gives the average values for the standard curing. The use of SAP resulted in some decrease in compressive strength. The same trends were obtained for compressive strength measured on halves of small beams with 40x40 mm cross-section (cf. Table 8.4), uniaxial tensile and flexural strength, except for tensile strength of oven-stored specimens.

Esteves et al. [11] tested mortar mixes at different w/c (from 0.25 to 0.35) under different curing conditions ranged from 30% to 100% of relative humidity at 20°C. They observed 20% reduction in compressive strength of SAP-containing mortars, when water curing was applied. At the same time, the results of testing mortars cured in adverse curing conditions were more favorable for SAP mixtures, which maintained their compressive strength. The authors concluded that mortars internally cured by means of SAP "seem to be insensitive to external relative humidity change in terms of mechanical resistance".

Mechtcherine et al. [12] reported about noticeable decrease in compressive strength for ultra high performance SAP-containing concretes (w/c=0.22) at early ages (1-7 days). However, the measurements at 28 days exhibited only a minor decrease in strength in cases when the addition of SAP and extra water were relatively small (0.3% SAP). A higher dosage of SAP (0.6%) and internal curing water led to a considerable decrease in strength.

Dudziak and Mechtcherine [13] observed a reduction of compressive strength of ultra high performance concrete (UHPC), but found "no pronounced negative effects

on compressive strength upon adding SAP, if a right amount of extra water is added for the purpose of internal curing". In another work of these authors, [14], a slight decrease in compressive strength of beams and cubes made of both fine-grained and coarse-grained concrete with SAP and tested at the ages of 1, 3, 7 and 28 days, relatively to the reference mixtures, was reported. The reduction of compressive strength in the range between 3% and 20% was measured at all testing ages and was more severe, when higher dosage of SAP and more extra water were used. This reduction was associated with the increase of capillary and gel porosity in the internally cured systems. It is worth to note that for the mixtures with addition of 2.5% of steel fibers (by volume) nearly no effect of the SAP addition on the values of compressive strength could be observed.

Wang et al. [15] observed also that compressive strength of concrete decreases with a higher content of SAP or entrained water.

The results obtained in the works [12,13] about a noticeable decrease of compressive strength at early ages, but only a minor decrease at 28 days, are quite consistent with the trend observed earlier in mortar and concrete internally cured by means of lightweight aggregates and IC agents, other than SAP. This trend is well-described in the literature. As reported in the State of the Art Report of RILEM TC-196 ICC [16], the increase of compressive (or flexural) strength of internally cured concrete, in comparison with regular concrete, was observed in concrete with lightweight aggregates [17–19] and in mortar with SAP or lightweight aggregates [20,21], especially at later ages. In these studies, the early age strength was generally lower than that in the reference mixes. For example, in the work of Weber and Reinhardt [17] addition of pre-wetted LWA to the mix produced concrete with compressive strength continuously increasing up to 1 year and insensitive to curing conditions. Zhutovsky et al. [19] found that incorporation of saturated lightweight aggregates had a detrimental effect on early-age (i.e. at the age of 1-7 days) strength of mixtures made at w/c=0.33. However, the 28-days strength of the internally cured mixes was close to the reference strength.

One of the possible reasons of the enhancement of compressive strength of concrete internally cured by means of SAP and other IC agents with the age is the improvement of hydration degree of cement grains, which is achieved by internal curing of concrete. In addition, as suggested by Reinhardt et al. [22], cement hydration products at later stages can grow inside the original grain boundaries of the SAP particles, contributing to the strength of the mature system. However, cement hydration products in the SAP pore were not observed directly.

Dudziak and Mechtcherine [14] state that the interpretation of the effect of SAP addition on the compressive strength is not straightforward. On the one hand, a reduction in the strength of the concrete matrix can be generally expected for the mixtures with SAP and extra water. This is a result of the formation of entraining pores, initially filled with curing water and subsequently dried out. Such voids effect the strength negatively. On the other hand, due to the reduction of the autogenous shrinkage of the cement paste, the internal stresses resulting from the hindrance of shrinkage deformations by stiff aggregates must be considerably lower in the case of specimens with SAP. This is positive from the standpoint of concrete strength.

8.3 Tensile Strength

The effect of SAP on tensile strength at different types of loading - uniaxial load, flexural (bending) load and splitting is reviewed in the current section. As will be shown later, the effect of SAP on tensile strength can be different than that on compressive strength. The possible explanation of this behavior will be suggested.

In the work referred before [10], Mechtcherine et al. studied tensile strength of both reference concrete made at w/c=0.25 and concrete with 0.4% of SAP (w_e/c=0.04). Uniaxial tensile strength was measured on special dog-bone shaped prisms with non-rotatable loading plates. The cross-section of the prisms was 40x24 mm, the length of the narrow part of the specimens was 80 mm. Flexural strength was measured on beams with 40x40 mm cross-section with a length of 160 mm and a span of 120 mm. It can be seen that similarly to the trends with compressive strength, the use of SAP resulted in a decrease of tensile strength, except for uniaxial tensile strength of oven-stored specimens (see Table 8.5).

Tensile splitting strength of concrete specimens (w/c=0.25) at ages of 7 and 28 days was determined by Igarashi and Watanabe in the work [7] (Table 8.6).

It can be seen that this trend is similar to that of compressive strength (see Table 8.2).

Dudziak and Mechtcherine [14] found a slight reduction of bending strength in ultra high performance mortars and concretes made at low w/c (0.22) and internally cured by means of SAP, similarly to compressive strength.

Esteves et al. [11] tested mortar mixes at different w/c (from 0.25 to 0.35) under different curing conditions ranged from 30% to 100% of relative humidity at 20°C. They observed severe reduction in tensile strength of SAP-containing mortars, when water curing was applied: approximately 30% (unfortunately, it was not reported what kind of tensile strength was measured). However, the results of testing mortars cured in adverse curing conditions were much more favorable for SAP mixes, which showed in some cases tensile strength at 28 days higher than that of reference materials.

Table 8.5 Uniaxial tensile and flexure strengths of concrete made with and without SAP and cured at different conditions [10]

	Uniaxial tensile strength, MPa		Flexural strength, MPa	
	20°C/28d	90°C/2d	20°C/28d	90°C/2d
Reference concrete	10.1	7.1	16.4	17.4
Concrete with SAP	8.2	7.6	12.3	12.0

Table 8.6 Tensile splitting strength of concrete with SAP at 7 and 28 days [7]

		Tensile splitting strength, MPa	
SAP, %	w_e/c	7 days	28 days
0	0	5.0	7.5
0.35	0.045	4.1	6.7
0.70	0.09	3.5	5.0

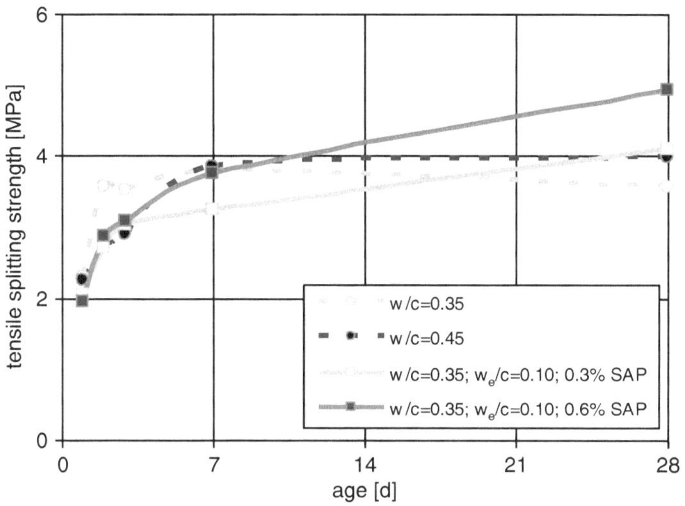

Fig. 8.4 Development of tensile splitting strength in time for concrete mixes containing SAP with two different contents, 0.3% and 0.6% (although made at the same content of entrained water – 10% by cement mass), in comparison with reference concrete mixes made at w/c=0.35 and 0.45. Adapted from [6]

Hoa Lam and Hooton [6] studied the effect of different materials having a potential to shrinkage mitigation, including SAP, on tensile splitting strength (Fig. 8.4). The specimens were cured under temperature of 23°C and 50% of air relative humidity. However, the results were different from those obtained under compression. While at early ages (3 and 7 days) the reference concrete made at w/c=0.35 was superior, like in compressive tests (see Fig. 8.1), SAP-containing mixture (with 0.6% SAP, because it seems that the mixture with 0.3% SAP contained too much entrained water) yielded the best result at 28 days, much better than even reference mixes. It is interesting that in between 7 and 28 days the strength gain for both SAP-containing concrete, with 0.3% and 0.6% of SAP, was the highest among all the mixes studied in this work.

It is known that tensile strength is often considered as a good indication of cracking resistance, and depends strongly on the existing microcracks, which can provoke the formation of macro-crack under tensile stresses. In view of this, the highest tensile strength of the mix with SAP at 28 days can be explained by the fact that SAP, in parallel to shrinkage mitigation, successfully increases the cracking resistance of the material.

8.4 Elastic Properties

Dudziak and Mechtcherine [14] found almost no difference between elasticity moduli of internally cured mortars and concretes, in comparison with those of reference ones. They tested ultra high performance mortar and concrete made at w/c=0.22, while the extra water was added (w_e/c=0.04) in the internally cured mixes. For instance, the reference ultra high performance mortar showed elasticity modulus of 48.8 GPa at 28 days, while the internally cured mix containing 0.3% SAP yielded 46.8 GPa. The similar results were obtained with steel fiber reinforced UHPC, when the reference mixture showed 54.7 GPa, and the SAP-containing mix gave 53.4 GPa. The standard deviation of these measurements did not exceed 1.3 GPa. As we can see, in both mortar and concrete only slight decrease of elasticity modulus was obtained, which was within the accuracy of the measurements.

8.5 Mechanical Properties of Concrete Made with SAP as Water Retaining Agent

Introduction of dry SAP into concrete mixture made of aluminate cement, without adding more mixing water, was studied by Gao et al. in the work [23] using statistical methods of experimental design. Three factors, amount of SAP (0.20, 0.35 and 0.50% by cement mass), particle size (20-40, 40-60 and >60 mesh) and water/cement ratio (0.32, 0.36 and 0.40), were investigated.

Adding SAP to aluminate cement paste prepared with a cement/water ratio of 0.40 increased the average compressive strength from 36.1 MPa (0% SAP) up to 40.5 MPa (0.2% SAP) and 44.4 MPa (0.6% SAP).

The average modulus of elasticity also increased: from 7.8 GPa (0% SAP) up to 10.1 GPa (0.2% SAP) and 11.1 GPa (0.6% SAP).

The average splitting tensile strength increased from 3.3 MPa (0% SAP) up to 5.8 MPa (0.2% SAP) and 7.0 MPa (0.6% SAP).

Adding SAP to concrete made on aluminate cement showed that the compressive strength of concrete increased linearly with decreasing mesh number of SAP, that is, increasing particle size and cement/water ratio, but parabolically with increasing amount of SAP and cement/water ratio. The modulus of elasticity showed the same trend, and the splitting tensile strength increased parabolically with decreasing mesh number of SAP, increasing amount of SAP and increasing cement/water content.

Larianovsky [24] studied two methods of using SAP in high-strength concrete, both as an agent for internal curing introduced into the mixtures together with the extra water required for internal curing (w_e/c=0.04), and also as a water-reducing agent (i.e. when the same SAP was introduced without adding extra water). This study has shown that whereas the compressive strength at the 1 day was reduced by SAP addition (when SAP was used as a water-retaining agent), the strength at 7 and 28 days slightly increased (Fig. 8.5 and Fig. 8.6). The improved strength at later ages may result from an increased degree of hydration caused by release of extra water from SAP. When SAP was used as an IC agent, the reduction of strength at age of 1 day was more severe, however at 7 and 28 days strength of SAP concrete was practically the same as that of the reference concrete.

Paiva et al [25] introduced 0.2%, 0.35% and 0.5% of SAP (by weight of cement) as retaining agent and showed that the presence of SAP almost did not change

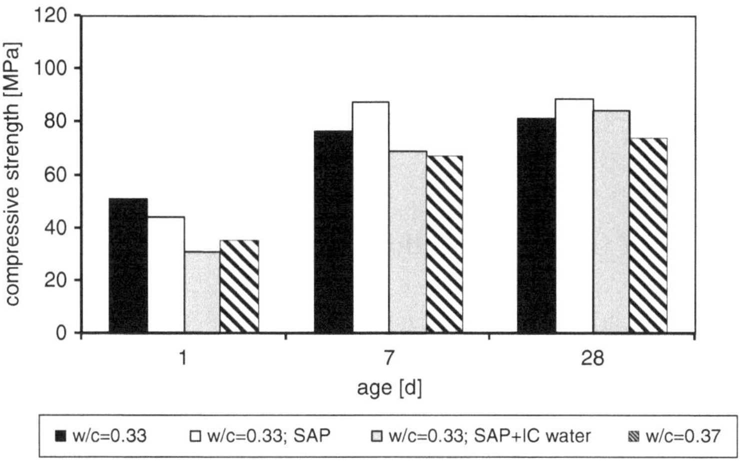

Fig. 8.5 Compressive strength of concretes made with and without SAP at w/c=0.33, in comparison with strength of concrete made at w/c=0.37 without SAP. Adapted from [24]

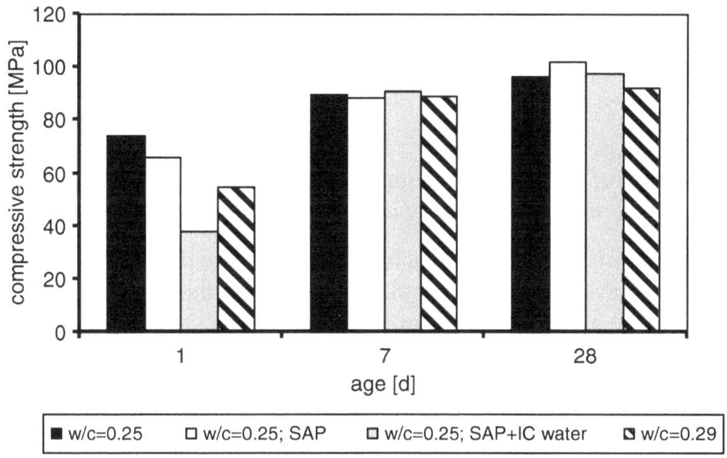

Fig. 8.6 Compressive strength of concretes made with and without SAP at w/c=0.25, in comparison with strength of concrete made at w/c=0.29 without SAP. Adapted from [24]

flexural strength in comparison with the reference mortar, but a small decrease of compressive strength occurred.

It has to be emphasized that using SAP as water retaining agent may not be favorable in terms of strength in cases when SAP is introduced in cement systems of too low w/c. Dudziak and Mechtcherine [14] tested ultra high-performance mortar made at w/c=0.22, both with 0.4% SAP and without SAP, and reported almost the same compressive strengths and elasticity moduli. The bending strength even decreased severely (by 30% at 28 days). The absence of the strength improvement and significant decrease of bending strength is not surprising and can be first of all attributed to the effect of voids induced by SAP addition in a very brittle matrix.

8.6 Effect of Curing Conditions

The correct comparison of the mechanical properties of concrete with and without SAP should take into account a possible effect of curing conditions. Most of the researchers use the sealed curing for both testing autogenous shrinkage and strength. However, some authors determine strength and other mechanical properties applying moist or air-drying curing, and it can influence the results. Unfortunately, there are not many publications, where different curing conditions are applied in the same study and compared. In view of this, it would be especially interesting to have a closer look into their results.

Usually, the researchers do not vary by the curing temperature, which is kept around 20°C (so called room temperature) till the test execution. However, the hygral conditions at curing can be different. We would like to group different curing conditions applied in the studies reviewed before onto the following categories:

- Sealed curing, or S;
- Moist curing, or M (air relative humidity RH is at least 95%);
- Drying curing, or D (RH is less than 95%).

Can the effect of curing conditions be significant enough? In order to answer this question, let us have a look at Table 8.7 summarizing the results of the following studies. This table shows the effect of SAP addition on strength determined under different curing conditions (S, M and D) at the age of 28 days only.

It can be seen that the most of the researchers report the strength reduction, however at sealed curing a few results demonstrate no influence of SAP addition or even slight increase in concrete strength.

Moist curing seems to be less favorable for the SAP-contained mixtures, then their exposure to drying in the open air. Esteves et al [10] state, in particular, that "water-entrained mortars maintain their strength despite of adverse curing conditions", remaining insensitive to the changes in relative humidity of the ambient air.

At the same time, we would not like to overestimate the effects of SAP addition on compressive strength and other mechanical properties, except shrinkage and crack resistance of concrete. Therefore, the main focus of using SAP in high-performance concrete should be done on studying and optimizing its potential in mitigating self-desiccation, autogenous shrinkage and cracking under restraint.

Table 8.7 Effect of SAP on concrete strength at 28 days under different curing conditions

Ref.	Curing conditions		
	Moist curing (M)	Drying curing (D)	Sealed curing (S)
[4]	Compressive strength decreased by 16-31%, depending on the SAP content		
[6]	Compressive strength decreased by 28-35%	Compressive strength decreased by 26-28%	
[7]			Compressive strength decreased by 3-22%; tensile strength decreased by 11-33%, depending on the SAP content
[8]	Compressive strength decreased by 8-14%		
[9]			10% reduction of compressive strength in pastes; no influence of SAP addition in mortars
[9]		Compressive strength decreased by 2% (cubes) and 13% (prisms); uniaxial tensile strength - by 19%; and flexural strength - by 25%	Compressive strength decreased by 10% (cubes) and 13% (prisms); uniaxial tensile strength increased by 7%; and flexural strength decreased by 31%
[10]	Compressive strength decreased by ~30%	Compressive strength decreased by ~20%	
[12]			Compressive strength decreased by 12-30% in prisms and by 4-36% in cubes, depending on the SAP content
[13]			Almost no compressive strength reduction and slight increase of flexural strength in SAP mixes
[24]			Slight increase (3-4%) in compressive strength

8.7 Summary and Conclusions

The main target of using SAP in high-performance concrete is to mitigate self-desiccation and autogenous shrinkage. A further goal would be not, or not substantially, to impair the concrete strength. In view of this, the efforts of researchers and technologists should be focused on these two simultaneous goals.

In reality, the introduction of SAP into concrete mixture results often in a certain strength reduction at early ages. This slight strength reduction is usually observed at the age of a few days. However, the correct dosages of both SAP and internal curing water and also optimum technological procedure (mixing, casting and curing) help to overcome this strength reduction. Some studies report that after 28 days the strength may even exceed that of the reference concrete mixture. One of the possible reasons of the enhancement of compressive strength of concrete containing SAP with the age is the improvement of hydration degree of cement grains, which is achieved by internal curing. In addition, cement hydration products at later stages grow inside the original grain boundaries of the SAP particles, contributing to the strength of the mature system.

Introduction of SAP as internal curing agent had a little influence on the strength of high performance mortars, while the strength of cement pastes was reduced more severely. The different effect of SAP introduction in cement pastes and mortars can be explained by different largest defect size present in the solid matrix of these materials.

The influence of SAP on tensile strength is not necessarily the same as that on compressive strength. There are indications that introduction of SAP in the system can improve tensile strength at mature ages, more than compressive strength. It can be related to the fact that tensile strength is sensitive to cracking, so any improvement of cracking resistance by means of self-desiccation/shrinkage mitigation is expected to be also beneficial in terms of tensile properties.

Strength of SAP-containing cementitious materials is also dependent on external curing conditions. The more severe strength reduction is usually observed under water curing. This type of curing is not indicative of the potential of using SAP. In contrary, using SAP seems to be more favorable under adverse external curing conditions, when air relative humidity is low. In this case SAP-containing systems at least maintain their compressive strength, while their tensile strength can even improve, compared with the reference materials.

The elasticity modulus of SAP-containing concrete is approximately the same as in reference concrete without SAP.

The use of SAP as water-retaining agent (without introducing additional water required for internal curing) helps to avoid the strength reduction in a very early age, and sometimes provides the strength at 28 days slightly higher than in the internally cured concrete. At the same time, this use of SAP does not enhance concrete strength significantly at mature ages, and therefore cannot compete with traditional polymeric water-reducing admixtures.

References

[1] Jensen OM, Hansen PF (2002) Water-entrained cement-based materials – II. Experimental observations. Cement and Concrete Research 32 (4): 973–978

[2] Bentz DP, Geiker M, Jensen OM (2002) On the mitigation of early age cracking. In: Proc. 3rd Int. Sem. on Self-desiccation and its Importance in Concrete Technology, Persson B and Fagerlund G (eds) Lund University, Lund, Sweden, 195–204

[3] Mönnig S (2005) Water saturated super-absorbent polymers used in high strength concrete Otto-Graf-Journal 16: 193–202
[4] Craeye B, De Schutter G (2006) Experimental evaluation of mitigation of autogenous shrinkage by means of a vertical dilatometer for concrete. In: RILEM Proc. PRO 52 Volume Changes of Hardening Concrete: Testing and Mitigation, Jensen OM, Lura P, Kovler K (eds) RILEM Publications S.A.R.L., Bagneux France, 21–30
[5] Craeye B, Geirnaert M, De Schutter G (2010) Super absorbing polymers as an internal curing agent for mitigation of early-age cracking of high-performance concrete bridge decks. Construction and Building Materials 25 (1): 1–13
[6] Hoa Lam, Hooton RD (2005) Effects of internal curing methods on restrained shrinkage and permeability. In: Proc. 4th Int. Sem. on Self-desiccation and Its Importance in Concrete Technology, Persson B, Bentz D, Nilsson LO (eds) Lund University, Lund, Sweden, 210–228
[7] Igarashi S-I, Watanabe A (2006) Experimental study on prevention of autogenous deformation by internal curing using super-absorbent polymer particles. In: RILEM Proc. PRO 52 Volume Changes of Hardening Concrete: Testing and Mitigation, Jensen OM, Lura P, Kovler K (eds) RILEM Publications S.A.R.L., Bagneux, France, 77–86
[8] Piérard J, Pollet V, Cauberg N (2006) Mitigating autogenous shrinkage in HPC by internal curing using superabsorbent polymers. In: RILEM Proc. PRO 52 Volume Changes of Hardening Concrete: Testing and Mitigation, Jensen OM, Lura P, Kovler K (eds) RILEM Publications S.A.R.L., Bagneux, France, 97–106
[9] Lura P, Durand F, Loukili A, Kovler K, Jensen OM (2006) Compressive strength of cement pastes and mortars with suberabsorbent polymers. In: RILEM Proc. PRO 52 Volume Changes of Hardening Concrete: Testing and Mitigation, Jensen OM, Lura P, Kovler K (eds) RILEM Publications S.A.R.L., Bagneux, France, 117–126
[10] Mechtcherine V, Dudziak L, Schulze J (2006) Internal curing by super absorbent polymers (SAP) – effects on material properties of self-compacting fibre-reinforced high performance concrete. In: RILEM Proc. PRO 52 Volume Changes of Hardening Concrete: Testing and Mitigation, Jensen OM, Lura P, Kovler K (eds) RILEM Publications S.A.R.L., Bagneux, France, 87–96
[11] Esteves LP, Cachim P, Ferreira VM (2007) Mechanical properties of cement mortars with superabsorbent polymers. Advances in Construction Materials, 451–462
[12] Mechtcherine V, Dudziak L, Hempel S (2009) Mitigating early age shrinkage of Ultra-High Performance Concrete by using Super Absorbent Polymers (SAP). In: Creep Shrinkage and Durability Mechanics of Concrete and Concrete Structures – CONCREEP-8, Tanabe T et al (eds) Taylor & Francis Group, London, UK, 847–853
[13] Dudziak L, Mechtcherine V (2008) Mitigation of volume changes of Ultra-High Performance Concrete (UHPC) by using Super Absorbent Polymers. In: Proc of the 2nd International Symposium on Ultra High Performance Concrete, Fehling E et al (eds) Kassel University Press GmbH, Germany, 425–432
[14] Dudziak L, Mechtcherine V (2010) Reducing the cracking potential of Ultra-High Performance Concrete by using Super Absorbent Polymers (SAP). In: Advances in Cement-Based Materials, van Zijl GPAG, Boshoff WP (eds) Taylor and Francis Group London UK, 11–19
[15] Wang F, Zhou Y, Peng B, Liu Z, Hu S (2009) Autogenous shrinkage of concrete with Super-Absorbent Polymer. ACI Materials Journal 106(2): 123–127
[16] Lura P, Jensen O, Igarashi S (2007) Chapter 6: Experimental methods to study internal water curing. In: RILEM Report 41 Internal Curing of Concrete, Kovler K, Jensen OM (eds) RILEM Publications S.A.R.L. Bagneux, France, 57–69.
[17] Weber S, Reinhardt HW (1997) A new generation of high performance concrete: concrete with autogenous curing, Adv Cem Based Mater 6 (2): 59–68
[18] Lura P (2003) Autogenous deformation and internal curing of concrete. PhD Thesis, Delft University of Technology, Delft, The Netherlands

[19] Zhutovsky S, Kovler K, Bentur A (2004) Influence of cement paste matrix properties on the autogenous curing of high-performance concrete. Cement and Concrete Composites 26 (5): 499–507
[20] Bentz DP, Geiker M, Jensen OM (2002) On the mitigation of early age cracking. In: Proc 3rd Int Sem on Self-desiccation and Its Importance in Concrete Technology, Persson B, Fagerlund G (eds) Lund University, Lund, Sweden, 195–204
[21] Lura P, Bentz DP, Lange DA, Kovler K, Bentur A (2004) Pumice aggregates for internal water curing. In: PRO 36: Proc Int RILEM Symp on Concrete Science and Engineering – A Tribute to Arnon Bentur, Northwestern University Evanston, Illinois, USA, Kovler K, Marchand J, Mindess S, Weiss J (eds) RILEM Publications S.A.R.L. Bagneux, France, 137–151
[22] Reinhardt HW, Assmann A, Mönnig S (2008) Superabsorbent polymers (SAPS) – an admixture to increase the durability of concrete. In: Proc 1st International Conference on Microstructure Related Durability of Cementitious Composites, Nanjing, China
[23] Gao D, Heimann RB, Alexander SDB (1997) Box-Behnken design applied to study the strengthening of aluminate concrete modified by a superabsorbent polymer/clay composite. Adv Chem Res 9: 93–97
[24] Larianovsky P (2007) Internal curing of concrete using super-absorbent polymers. MSc Thesis, Technion – Israel Institute of Technology, Haifa
[25] Paiva H, Esteves LP, Cachim PB, Ferreira VM (2009) Rheology and hardened properties of single-coat render mortars with different types of water retaining agents. Construction and Building Materials 23: 1141–1146

Chapter 9
Effect of Superabsorbent Polymers on Durability of Concrete

Hans-Wolf Reinhardt and Alexander Assmann

Abstract The chapter deals with various aspects of durability influenced by SAP. Although the compressive strength is reduced with the addition of SAP it could be shown that the water and oxygen permeability remain the same or are improved. Due to the generation of air-filled pores by SAP the frost resistance with deicing salts is considerably superior to reference concrete without SAP. It is comparable to concrete with airentraining agents. Finally, chloride migration is also reduced depending on the type of SAP.

9.1 Introduction

Durability means resistance against environmental influences such as frost action, ingress of chlorides, ingress of alkalis, ingress of carbon dioxide, acid attack, attack by pure water, wear, high and low temperatures, and other attacks. The attack can take place at the surface of a concrete structure or it can develop in the interior. The transport of fluids and gases is necessary for the deterioration of concrete structures, i.e. for corrosion of reinforcement, for alkali-silica reaction, for leaching, for frost and deicing-salt damage. To assess the resistance of concrete against the various attacks it is of paramount interest to know the transport properties of concrete. In the following, the knowledge on gas and water permeability, capillary suction, frost resistance, and chloride migration on SAP modified concrete is being reviewed and evaluated.

H.-W. Reinhardt (✉) • A. Assmann
Department of Construction Materials, University of Stuttgart, Stuttgart, Germany
e-mail: reinhardt@iwb.uni-stuttgart.de

V. Mechtcherine and H.-W. Reinhardt (eds.), *Application of Superabsorbent Polymers (SAP) in Concrete Construction*, RILEM State of the Art Reports 2,
DOI 10.1007/978-94-007-2733-5_9, © RILEM 2012

9.2 Permeability

The literature does not contain many references on permeability. A test series has been reported from the University of Stuttgart in which the oxygen and water permeability and the capillary suction has been investigated [1]. The concrete mixtures were all the same.

9.2.1 *Mixture composition*

The used SAP polymer was an acrylic type which can be divided into three classes of particle diameter in dry condition which are < 63 μm (s), 63 to 125 μm (m), and 125 to 250 μm (l). The water absorption of the particles was determined to be 24 g water/ g polymer which makes the particle diameters to swell by a factor of 3.26. From the added mass of SAP and the particle distribution, the total SAP pore volume and the total number of pores in 1cm³ concrete can be estimated. The values are listed in Table 9.1.

Three reference mixtures with w/c of 0.50 (M1), 0.42 (M2) and 0.36 (M3) and five mixtures with SAP addition (M12s, M12m, M12l, M13, M23), varying SAP content and particle size distribution, were produced. To evaluate the influence of SAP on concrete properties, for each mixture with SAP addition two reference mixtures can be taken for comparison: one reference mixture with same w/c-ratio and a second with w/c according to w/c_{eff} of the SAP-mixture. w/c_{eff} is defined as w/c-ratio by disregarding the water stored in SAP pores. Table 9.1 shows also the mixture composition in detail.

The used cement type was a CEM I 42.5 R with a constant amount of 450 kg/m³. The chosen grading curve of Rhine gravel aggregates was an approximated 'AB' curve according to DIN 1045-2 with a maximum grain size of 8 mm in diameter. To achieve the same consistency for all mixtures, superplastizicer Glenium ACE 30 was used.

9.2.2 *Properties of fresh concrete*

The total mixing time was 5 min 30 s by normal speed in a pan-type mixer with a capacity of 150 dm³. After mixing aggregates, cement and SAP in dry condition for the first 60 s - to achieve a better dispersal of SAP - half of the water was added. Within the following 1 min 30 s, the water absorption of SAP started. Superplasticizer and remaining water were added and the mixer was mixing for another 3 min. To achieve higher workability of M23, 0.3% superplasticizer by mass of cement was added supplementary. Therefore the total mixing time increased to 7 min 30 s.

Spread, temperature, air content and bulk density of the fresh concretes were measured directly after mixing. The target value of concrete spread was 45 cm. Actual values range between 36 and 53 cm. The measured air content is correlated with the spread: a more flowable concrete means better compactibility, and consequently a lower air content. The air content range was 0.70 to 3.40% by volume.

Table 9.1 Concrete mixtures produced for the experimental series on permeability

Mixture			M1	M2	M12s	M12m	M121	M3	M13	M23
Water	W	l/m^3	225	189	225	225	225	162	225	189
	w_{stored}	l/m^3			36	36	36		63	27
	$(w/c)_{tot}$		0.50	0.42	0.50	0.50	0.50	0.36	0.50	0.42
	$(w/c)_{eff}$		0.50	0.42	0.42	0.42	0.42	0.36	0.36	0.36
NWA	0 – 1.2 mm	kg/m^3	480	509	480	480	480	530	479	509
	1.2 – 2 mm	kg/m^3	321	339	320	320	320	354	320	339
	2 – 4 mm	kg/m^3	321	339	320	320	320	354	320	339
	4 – 8 mm	kg/m^3	481	511	480	480	480	530	479	509
	Total	kg/m^3	1603	1697	1600	1600	1600	1768	1598	1695
Admixtures										
SAP	f_{dry}	kg/m^3			1.50	1.50	1.50		2.63	1.13
	Fraction	m-% b.c			0.33	0.33	0.33		0.58	0.25
	Size	mm			< 63	63 – 125	125 – 250		63 – 125	63 – 125
	Absorption	g/g			24	24	24		24	24
	Median pore radius r_{sm}	mm			65	147	244		147	147
	Pore volume	% b. V.			3.75	3.75	3.75		6.6	2.8
	Total number of pores	p. cm^3			120170	4970	785		8745	3710
Superplasticizer	Amount	kg/m^3	0.68	1.80	1.80	2.03	2.03	5.63	4.05	3.60
type PCE	Fraction	mÐ% b. c	0.15	0.40	0.40	0.45	0.45	1.25	0.90	0.80

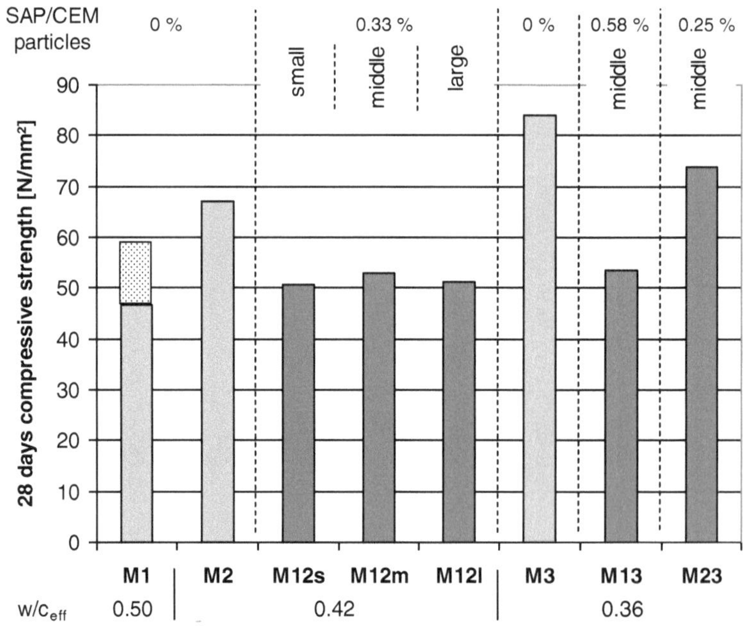

Fig. 9.1 28 d compressive strength f_c [N/mm²]

9.2.3 Storage conditions

All specimens were left in their moulds for 24 hours at 20°C and 100 % RH. To simulate different curing periods, half of the specimens were stored for three more days at 20°C and 100 % RH. The other half was put at 23°C and 50 % RH. Specimens used for capillary suction, water and oxygen permeability were sealed at the perimeter with epoxy one week after mixing. The tests were performed at an age of 28 days. Specimens used for oxygen permeability were used later for mercury intrusion porosimetry.

9.2.4 Compressive strength

Three cubes with 150 mm edge length were tested at an age of 28 days. The average values of measured compressive strength f_c in N/mm² are given in Figure 9.1. The standard deviation ranges between 0.5 N/mm² (M13) and 2.7 N/mm² (M1).

In this series, strength of SAP modified concrete with low SAP content tends towards strength of reference concretes with the same w/c. The results of M12s, M12m, M12l and M13 verify this hypothesis: all SAP mixtures with w/c = 0.50 and SAP pore volumes of 3.7 % to 6.50 % by total volume, show almost same values for f_c. The difference in pore size as well as in number of pores (M12s, M12m, M12l) doesn't seem to have an effect, because of little polymer pore volume compared to

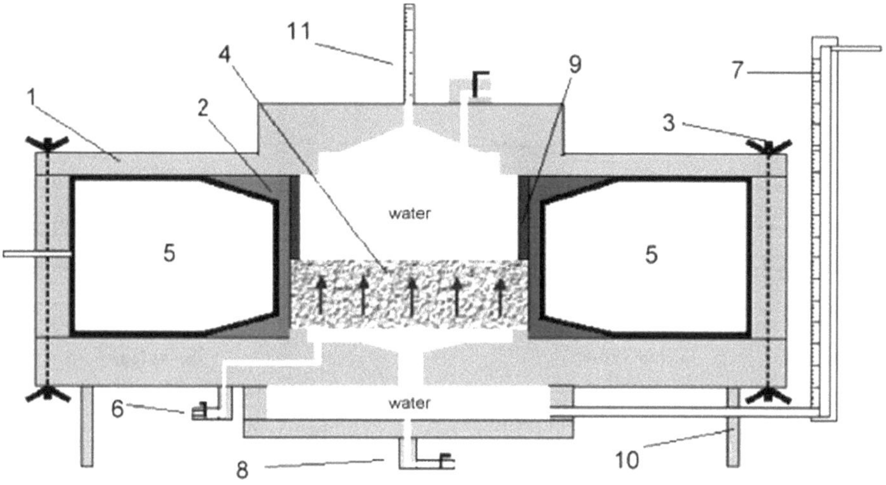

1. aluminium housing 2. seal pressure sleeve 3. fixing screw 4. concrete sample
5. pressure tube 6. deariation valve 7. measuring capillary (water pressure)
8. water inlet 9. backup ring 10. support 11. measuring capillary (flow rate)

Fig. 9.2 Test cell for water permeability

volume of air pores. The reference mixture M1 shows a higher value for f_c than expected. Due to this result which could not explained, 3 more cubes were cast in a second batch with 2.9 % air content, 44 cm spread and a fresh concrete temperature of 18.2°C. Their 28 days f_c reached only 46.6 N/mm², which is an acceptable value.

9.2.5 Permeability testing procedure

A cut through the cell for water permeability experiments is shown in Figure 9.2. Its leak-tightness was secured by the rubber tube (5) with a pressure of 11 bar. The cell for oxygen permeability is analogous [2].

Gas flow through porous materials like concrete can be described by Hagen-Poiseuille's law:

$$Q = K \cdot \frac{A}{\eta \cdot L} \cdot \frac{\left(p_i^2 - p_o^2\right)}{2p} \tag{9.1}$$

Respectively:

$$K = \frac{Q \cdot \eta \cdot L}{A} \cdot \frac{2p}{\left(p_i^2 - p_o^2\right)} \tag{9.2}$$

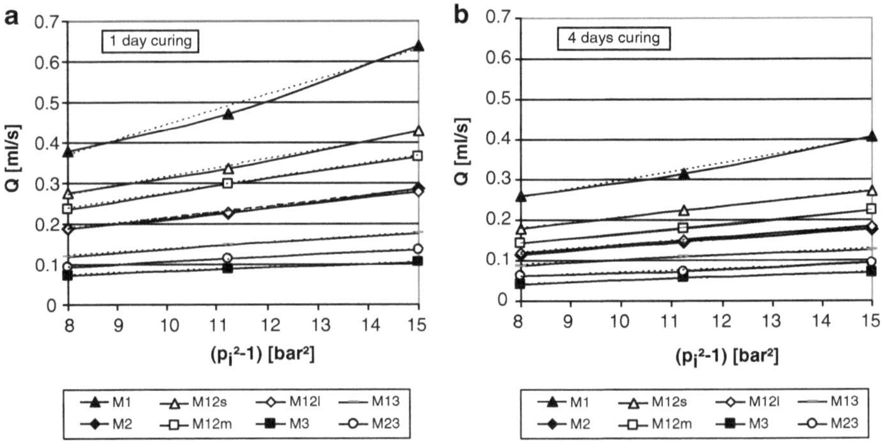

Fig. 9.3 (**a**) Oxygen flow rates of specimens with 1 day curing time in dependence on inlet pressure (p_i^2-1); (**b**) Oxygen flow rates of specimens with 4 days curing time in dependence on inlet pressure (p_i^2-1)

For oxygen applies:

Q	[cm³/s]	Flow rate
K	[m²]	Specific permeability coefficient
A	[mm²]	Test surface
L	[mm]	Height of specimen
$p_o = p$	[10^{-5} N/mm²]	Outlet and atmospheric pressure
p_i	[10^{-5} N/mm²]	Inlet pressure
η	[10^{-5} Ns/m²]	Viscosity of test gas

Oxygen was chosen as testing gas: First of all, it causes no change of microstructure like CO_2 for example. Second, it is the oxidizing agent in corrosion processes of reinforcement and therefore relevant for durability and third it can be handled safely in the laboratory.

Flow rates Q of oxygen through cylindrical specimen with diameters of 100 mm and 50 mm in height were determined by a bubble counter. Before cutting the specimen at an age of 7 days, the perimeter was sealed with epoxy to eliminate gas flow due to shrinkage cracks. At the end of preparation, the testing surfaces were polished.

At 20°C room temperature, the viscosity of oxygen is approximately $\eta = 2.02 \cdot 10^{-5}$ Ns/m² and the atmospheric pressure is nearly 1 bar. Flow rates were measured for three inlet pressures: 3.0, 3.5 and 4.0 bar. To reach constant values for Q, oxygen flows were not measured until 5 min up to 30 min after the specimens had been put under the adjusted inlet pressure. For laminar flow, Q has to be proportional to (p_i^2-1). The specific permeability coefficient K can be regarded as gradient of the graphs in Figures 9.3a and 9.3b. Each measuring point is the average of two tested specimens. The dotted lines represent the best fitting regression lines. As the results show, the assumption of a laminar flow is acceptable.

The water permeability was determined on specimens with 30 mm in height and 150 mm in diameter. They were put in the cell and fixed with a metal ring towards the top cover. The whole cell was closed up airtight by pressure tube and filled with water. To apply a water pressure of 7 bar, nitrogen gas was used as a buffer. The water flow through specimen was read on measuring capillaries. Before the measurements started, the specimen in the proof cell was under water pressure of 7 bar for 24 hours to make sure that it is completely water saturated. After these 24 hours the preparations had been finished and the measurements were started. Flow rates were determined after 1 hour and 24 hours, according to [1]. Specimens with 1 day curing time provided good results.

In case of laminar flow, Eqs. (1) and (2) are valid for water permeability, too. From the fact that water can be seen as incompressible, it follows:

$$p = \frac{p_i + p_o}{2} \tag{9.3}$$

The water permeability coefficient k_w can be calculated by converting equation (1) to:

$$k_W = \frac{Q \cdot L}{A \cdot \Delta h} \tag{9.4}$$

In particular:

k_w	[m/s]	Water permeability coefficient
Q	[m³/s]	Flow rate
A	[m²]	Test surface
L	[m]	Height of specimen
Δh	[m]	Water pressure

9.3 Oxygen Permeability

The measured oxygen permeability coefficients are presented in Figure 9.4. They have coefficients of variation between 6 and 13 %. In all cases, an extension of curing time leads to a reduction of permeability: 4 d show smaller values of K than 1 d. Average values of 4 days cured specimens range between 60 and 70 % of 1 day cured specimens.

As SAP is concerned a look at mixtures M12s, M12m and M12l is conspicuous, that an increasing particle size distribution of SAP causes less permeability. Indeed the total number of pores is less by same total pore volume, but large SAP pores seem to generate a greater density of microstructure than small SAP pores do. The most notable effect could be that smaller numbers of SAP pores do not create a continuous structure. Furthermore, the volume of a spherical pore depends on the third power of the diameter whereas the surface of a sphere grows only with the second power. Therefore, the larger the pores with the same total porosity the less accessible towards oxygen and other fluids. Another speculation is that larger pores become denser because of their better water supply

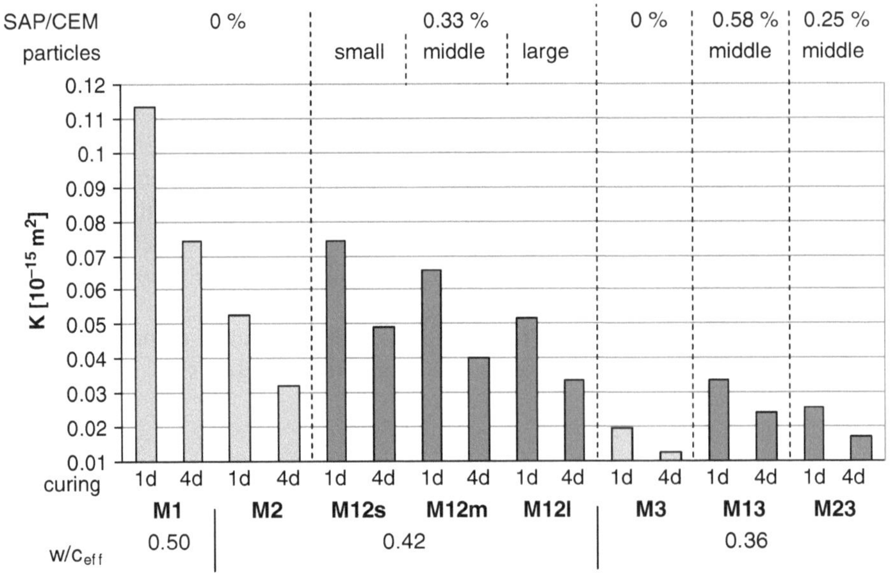

Fig. 9.4 Oxygen permeability coefficients [10^{-15} m²]

to the surrounding cement paste, accompanied by a stronger growth of hydration products into the pores.

In general, the determined permeability coefficients agree with values given by literature [3]. Beside the strength class, K highly depends on moisture content of tested specimens [4]. In [5] specimens of same mixtures were exposed to different climates. Higher moisture contents caused a decrease of K up to one or two orders of magnitude. To compare different concretes with each other, same storage conditions have to be kept. In our case, all specimens were dried in the oven by 60°C till they reached constant weight. If standard climate of 20°C and 65 % for example would have been kept, it could be anticipated that SAP mixtures would have shown lower values of K because of delayed desiccation in comparison to reference mixtures.

9.4 Water Permeability

The outcome of the measured water flow rates after 1 hour respectively 24 hours, leads to the values for k_w presented in Figure 9.5.

It can be seen clearly, that k_w decreases with increasing time. Some reasons are the reactivity of water with unhydrated cement particles and the swelling of the concrete itself. This causes further density of the matrix. The results of the water permeability coefficients after 24 hours confirm the tendency deduced from oxygen permeability measurements: increasing particle size of SAP causes less permeability.

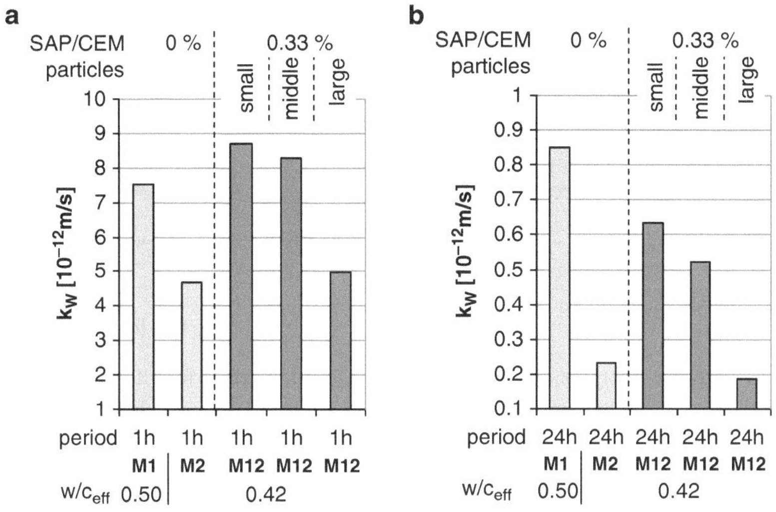

Fig. 9.5 (**a**) Water permeability coefficients after 1 hour; (**b**) Water permeability coefficients after 24 hours

However values of k_w determined after 1 hour show high variations. It seems likely, that the specimens were not saturated completely at the beginning of the tests. This assumption gets affirmed by another observation: air bubbles rose up at the non-pressurized surface during the measurements. On the one hand they might be coming from the dissolved pressure gas, on the other hand they might have emerged from the air in the larger pores which is pressed out. Both reasons weaken the value of the results. To get reliable data, the investigation about water permeability should be extended.

9.5 Capillary Suction

In case of water absorption of dry concrete samples, capillary suction is the most prominent transport mechanism. Cylindrical specimens with 150 mm in diameter were sealed with epoxy on the perimeter and cut on length. The cut surface was polished till all specimens had the same weight of 5000 g. Figures 9.6a and 9.6b show the absolute mass of absorbed water in g as well as the relative water absorption in % by mass over a period of 3 days.

An extension of curing time causes only a small decrease of water absorption by capillary suction. In general, the results of the SAP mixtures show the same tendency known from permeability testing: added SAP cause densification of the matrix and lead to less capillary porosity. High numbers of small particles create a coherent pore structure which absorbs water easier than a structure caused by a small number of larger particles. However, the difference between M12m and M12l

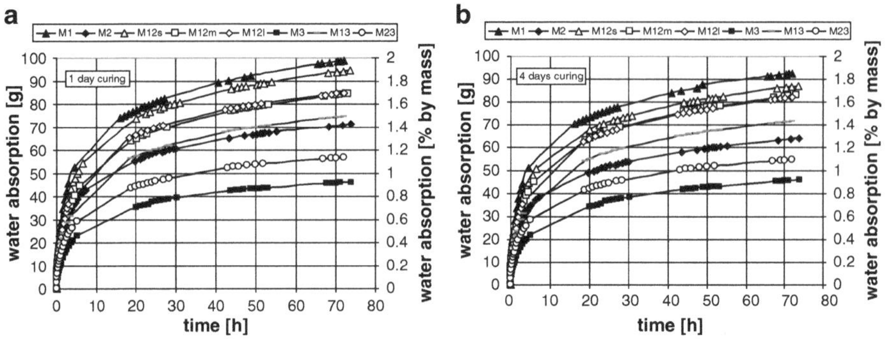

Fig. 9.6 (**a**) Water absorption by capillary suction as function of time for cylinders with 1 d curing time; (**b**) Water absorption by capillary suction as function of time for cylinders with 4 d curing time

is less pronounced than expected. Both mixtures show nearly the same performance. This finding refers to the fact that capillary pores are scaled below SAP pores. For capillary pores, which have diameters of 10^{-5} to 0.1 mm, capillary rise is inversely proportional to the pore radius. With growing pore sizes (> 0.1 mm), capillary action decreases till the influence of gravity prevails. Consequently high sized fractions of SAP pores are not relevant for capillary suction. Anyway, SAP pores get filled with water and have a share in the increase of mass. Different moisture contents in the specimens could have influenced the results, too. Due to the epoxy sealing at the perimeter and the large size of the samples, a period of 21 days at 60°C was not enough to dry the specimens out completely.

The graphs of water absorption get flattened over the time. The reasons are the same as mentioned with water permeability: water reacts with unhydrated cement particles and causes further densification of the matrix. Secondly the concrete swells when getting in contact with water [6]. Finally the gravity interferes with capillary action.

According to theory, the capillary uptake follows a square root of time law except for water at longer duration as stated above. So the water-absorption coefficient w has the unit $kg/m^2\ h^{1/2}$. Picking values of water absorption in kg/m^2 in dependence on square root of time over a period of the first 4 hours, a linear relationship appears (see Figure 9.7). Proper values for w [$kg/m^2\ h^{1/2}$] are given in Table 9.2.

It seems to be evident, that the presented results are subject to variance because the experiments have to get by with only one sample of each mixture. However, the rate of water absorption during the first 4 hours is almost independent of curing time. With Table 9.2 in mind, this phenomenon may be explained: measured median pore radius, which acts as an indicator for capillary pore size, does not change with an extended curing time of 3 days. The reason why mixture M12m has a smaller water absorption rate compared with M12l at the beginning, whereas both mixtures adopt same level at the end of the test, is unknown.

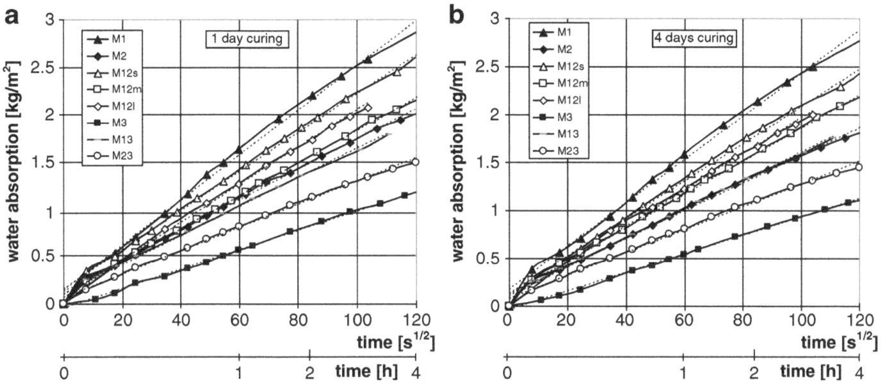

Fig. 9.7 (**a**) Water absorption by capillary suction vs. square root of time for cylinders with 1 d curing time; (**b**) Water absorption by capillary suction vs. square root of time for cylinders with 4 d curing time

9.6 Summary of Transport Properties and Porosity

Table 9.2 shows the results of the transport experiments and of mercury intrusion porosimetry (MIP). The results are dependent on the storage duration (1 d vs. 4 d).

The samples taken for mercury intrusion porosimetry had been used before to determine oxygen permeability. Thus both examinations were carried out with identical material. MIP generates the pore size distribution of tested material, but only pore size radii between 3.75 nm and 7500 nm are covered [7]. This is why the median pore radius pictures a factor for capillary pore sizes. SAP pores do not appear in pore size distribution determined by MIP, but nevertheless they are captured in total pore volume. The total pore volume given in % can be sectioned into three different types: air voids, SAP pores and capillary pores. The volume of capillary pores remains by subtracting air voids, measured in fresh concrete, and calculated SAP pores from total pore volume. In doing so, SAP pores are assumed to keep their size during hydration processes of the surrounding paste. Indeed this is a simplification.

Figure 9.8 shows the fractional pore volume. The histogram provides a better demonstration of the results.

By having a closer look at the capillary pore volume, it can be seen, that the calculated water absorption of SAP operated well: Mixtures with same w/c_{eff} show nearly same capillary porosity. With higher w/c, capillary porosity is increasing. Fully hydrated cement paste with $w/c < 0.42$ has a very low capillary porosity. Given that α is low, capillary porosity is much higher. Another evidence for the efficient water absorption is the indicated median pore radius: Mixtures with same w/c_{eff} show once more conformable values. Generally speaking, median pore radii decrease with decreasing w/c.

Specimens with 4 days curing time should show less capillary porosity which would be in good agreement with the higher degrees of hydration, this cannot be

Table 9.2 Summary of testing results of transport properties and porosity

Mixture		M1		M2		M12s		M12m		M121		M3		M13		M23	
Curing time		1 d	4 d	1 d	4 d	1 d	4 d	1 d	4 d	1 d	4 d	1 d	4 d	1 d	4 d	1 d	4 d
Oxygen permeability																	
K	$[10^{-17}m^2]$	11.35	7.42	5.25	3.21	7.44	4.93	6.56	4.02	5.17	3.38	1.96	1.24	3.38	2.40	2.58	1.71
Standard deviation	[%]	6.0	9.6	11.3	9.5	9.8	10.3	9.5	8.9	11.8	10.0	12.7	8.5	11.9	11.9	12.9	10.6
Capillary suction																	
w	$[kg/m^2h^{1/2}]$	1.42	1.35	0.97	0.87	1.23	1.17	1.04	1.04	1.18	1.14	0.61	0.58	0.92	0.91	0.74	0.71
Mercury intrusion (MIP)																	
Total pore volume	$[mm^3/g]$	76.1	71.2	55.2	57.0	76.3	74.6	79.3	70.4	74.4	76.5	45.6	44.2	79.7	74.3	55.4	58.5
Total pore volume	[% by vol.]	16.1	15.3	12.3	12.6	16.3	15.8	16.7	15.1	15.8	16.2	10.6	10.2	16.9	15.8	12.4	13.0
Median pore radius	[μm]	0.075	0.075	0.061	0.062	0.062	0.062	0.061	0.062	0.062	0.062	0.051	0.051	0.051	0.051	0.051	0.051
SAP pores & air voids																	
SAP pore volume*	[%]	-		-		3.75		3.75		3.75				6.60		2.80	
Air content	[%]	2.0		3.4		3.0		2.4		2.5		0.7		1.3		0.8	

*compare Table 9.1

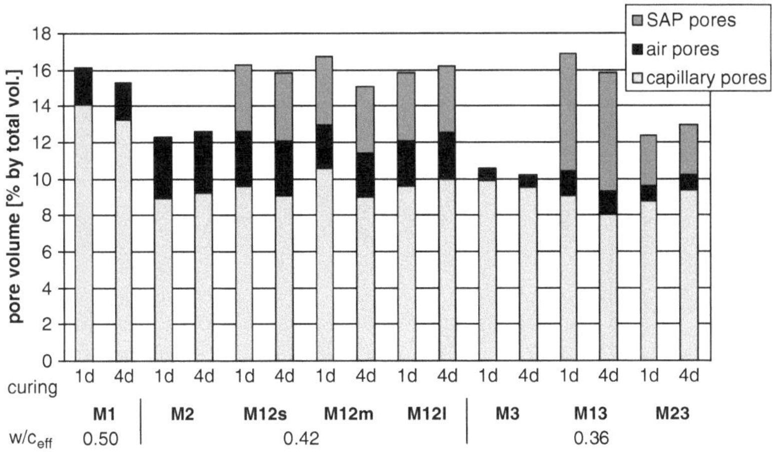

Fig. 9.8 Fractioned pore volume [% by vol]

seen at all results. An explanation for this could be the effect of a heat treatment while storing in the oven by 60°C. Hydration processes do not stop before relative moisture in the microstructure declines ca. 80 to 90 %, but pass off accelerated when exposed to higher temperature [8]. Both requirements might have been fulfilled for a short time at specific points in the sample and have distorted the results.

9.7 Freeze-Thaw Resistance

9.7.1 Mixtures

Eight mixtures were tested during this testing program (Table 9.3). Two different types of SAPs, one with high water absorption capacity, approximately 18 g/g, and one with smaller water absorption capacity, approximately 11 g/g were used for CDF testing. Two mixtures with air entraining agents (AEA) were also used for comparison.

9.7.2 Experimental methods

The frost tests were carried out with the CDF test (Capillary suction, Deicing solution, Freezing) acc. to RILEM [9]. After 24 h the specimens were demoulded and thereafter stored for 6 days at 20 ± 2°C and 100 % RH. After this period the cubes

Table 9.3 Concrete mixtures for frost testing

Designation	Unit	Ref 042	Ref 048	SAP I	SAP II	SAP III	SAP IV	LP74_5	LP 74_10
Cement	kg/m³	356	353	353	353	353	356	353	353
$Water_{(added)}$ [1)]	L	145.6	165.1	148.9	149.3	149.9	166.1	165.4	165.1
$Water_{(stored\ by\ SAPs)}$	L			15.4	14.1	14.1	15.5		
SAP	kg/m³			1.397	0.784	0.784	1.410		
Type				SAPD	SAPB	SAPB	SAPD		
Size	mm			0–250	63–125	< 63	0–250		
Absorption	g/g			11	18	18	11		
Superplasticizer	kg/m³	1.72	1.19	3.00	4.00	3.00	1.18	1.25	1.17
Air entraining agent	kg/m³							0.125	0.250
Aggregates	kg/m³	1879	1818	1830	1818	1818	1787	1727	1675
$w/c_{(total)}$ [2)]	-	0.42	0.48	0.48	0.48	0.48	0.51	0.48	0.48
$w/c_{(after\ absorption)}$ [3)]	-			0.42	0.42	0.42	0.47		
Air content [4)]	%	1.6	2.3	4.0	0.5	1.0	5.4	4.5	10.5

1)Total water added during mixing
2)Including water from the superplasticizer suspension and aggregate moisture
3)Absorption capacity of the SAP, determined according to [10], in g/g
4)Air content in fresh concrete

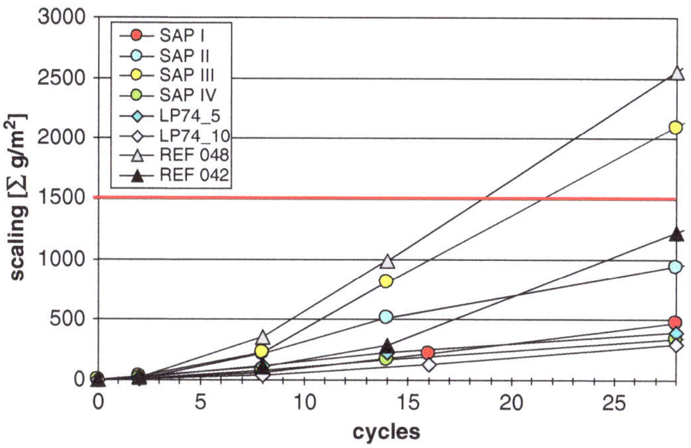

Fig. 9.9 Deterioration of the tested specimens after 28 cycles

were sawn into two halves by wet sawing, followed by a 21 days storage at 20 ± 2°C and 65 ± 5 % RH. Within this period the lateral sides of the cube halves were sealed with an epoxy resin. After 28 days they were stored in a water bath for another 7 days. The water had a NaCl content of 3 % by mass. The test surface had a size of 15 × 15 cm². Five cubes of each mixture were used for the freeze-thaw test. A freeze-thaw cycle lasted 12 hours. Within the first four hours the temperature dropped from 20°C to -20°C. The temperature stayed constant for another three hours and then it increased to 20°C. All examinations and the replacement of the water bath were completed within one hour, after which the next cycle started. The weighing of the scaled particles was done after 2, 6, 14, 28 and 56 cycles.

9.7.3 Scaling

As an experimental result, the scaling of concrete has been measured after 2, 6, 14, 28 and 56 cycles, i.e. after 1, 3, 7, 14 and 28 days of freeze-thaw testing. The CDF test is generally conducted until the 14th day. Some of the mixtures showed a very small increase of damage and therefore it was decided to extend the period by another test after 56 cycles. Figures 9.9 and 9.10 show the scaling of the mixtures after 28 and 56 cycles. The limit for frost/deicing resistant concrete is 1500 g/cm² after 28 cycle acc. to RILEM.

Figure 9.9 shows that two mixtures are above the 1500 g/m² line after 28 cycles, these are the reference mixture with water-cement ratio of 0.48 and SAP III mix with total water to cement ratio of 0.48 and grain size of SAP < 63 mm. All other mixtures stayed below the 1500 g/m² line. However, the development between the 28th and 56th cycle (Figure 9.10) showed interesting differences: while the SAP and the AEA mixtures had only a small increase of scaling and stayed well below the 1500 g/m² limit, the reference mixtures and the SAP III mixture deteriorated rapidly.

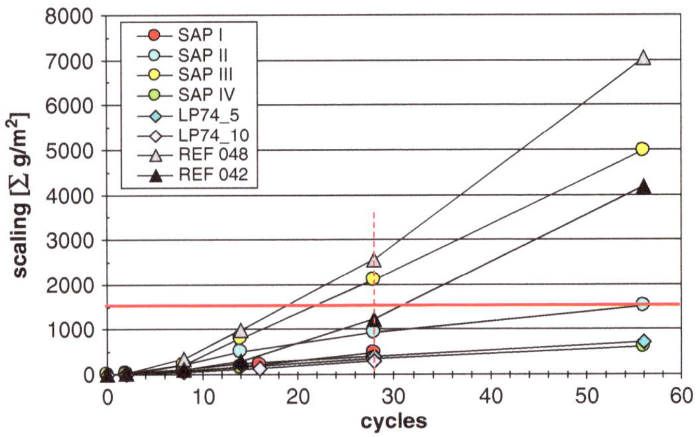

Fig. 9.10 Deterioration of the tested specimens after 56 cycles

Table 9.4 Calculated void spacing for the SAP mixtures

Mixture	Paste volume [l]	Particle number [N]	$VS_{cal.}$ [mm]	$VS_{meas.}$	Factor $VS_{meas.}/VS_{cal.}$
SAP I	280	$7.34 \cdot 10^9$	0.232	0.17	0.73
SAP II	281	$3.68 \cdot 10^9$	0.275	0.17	0.61
SAP III	281	$117.3 \cdot 10^9$	0.086	0.29	3.37
SAP IV	298	$7.41 \cdot 10^9$	0.237	-	-
REF 042	280	-	-	0.250	-

The void spacing is a good indication for the SAP efficiency. According to Powers [11] the void spacing (VS) is a good indication for the resistance of the cement paste against freeze-thaw damage. However, the determination of the VS is a difficult task and cannot be done without any laboratory tests. The SAPs provide an excellent possibility to estimate the VS without any advancing experiments. It can be done simply by calculating the approximate number of particles and their average distance to each other based on the assumption of saturation and equal spacing. This has been calculated in [12].

Table 9.4 shows the results for the examined SAP mixtures. Mixtures I and IV show comparable VSs and their scaling was also comparable to each other. The smaller size of the SAPs used for mixture III resulted in a very high amount of particles which decreased the distance between the particles almost to zero. The mixture showed very fast water absorption and a thick layer of water on the top of the specimens indicating a high permeability of the cement paste. The specimens showed scaling on the bottom and on the top resulting in the very high total scaling.

Table 9.5 Material data and composition of 1 m³ concrete used [13]

Material	Size [mm]	Absorption [%]	Density [kg/m³]	Mass [kg]	Volume [m³]
Cement CEM I 52.5	-	-	3150	390	0.124
Tap water	-	-	1000	164	0.164
Entrapped air (estimate)	-	-	-	-	0.015
Sea sand	0-4	0.5	2601	774	0.297
Sea gravel	4-8	0.6	2642	333	0.126
Sea gravel	8-16	0.5	2637	722	0.274

9.7.4 Influence of particle size

It is well known that the pores in a concrete matrix have to have a certain size and distance in order to ascertain frost resistance. As possible air-entraining measure SAP is used. The SAP of this investigation was a covalently crosslinked acrylamide/acrylic acid copolymer [13]. The concrete had the composition according to Table 9.5.

The dry SAP particles with a density of approximately 1500 kg/m³ were divided into the fractions 38 to 63 μm and 90 to 125 μm by sieving. During mixing of the concrete SAP absorbs approximately 12.5 g water per g dry SAP. The dry particles swell by water absorption and the two fractions of SAP got average diameters of 150 and 300 μm respectively. The total volume of SAP pores were adjusted to 1, 2, 3, 5 and 10 % with 150 μm SAP and 5, 10, 15, 25 and 35 % with 300 μm SAP relative to the volume of cement paste.

The freeze-thaw tests were carried out acc. to the European standard CEN/TS 12390-9 [14] where temperature cycles of -20 to +20°C during 24 hours apply. The test is continued up to the end of the 56th cycle. Figure 9.11 shows the amount of scaled material as a function of numbers of cycles.

It can clearly be seen that the SAP modified concretes behave much better than the reference concrete. Figure 9.12 shows the mean values of scaled material after 56 cycles as function of pore volume and spacing factor. The spacing factor was estimated to be in the range of 0.15 to 0.40 mm depending on the number of SAP pores.

It can be seen that scaling is reduced with higher amount of SAP pores. The Swedish standard SS 137244 [15] defines how to classify concrete after the frost test. "Good" performance is achieved when scaling amounts to 0.2 kg/m² and "very good" when it is only 0.1 kg/m². Most SAP modified concretes range under "good" and three of them under "very good". All reference concretes are outside the good and very good performance. As the pore size is concerned the larger SAP particles behaved the best.

9.7.5 Effect of SAP on freeze-thaw resistance of SHCC

Brüdern and Mechtcherine [16] investigated the multipurpose use of superabsorbent polymers (SAP) in strain-hardening cement-based composites (SHCC). While

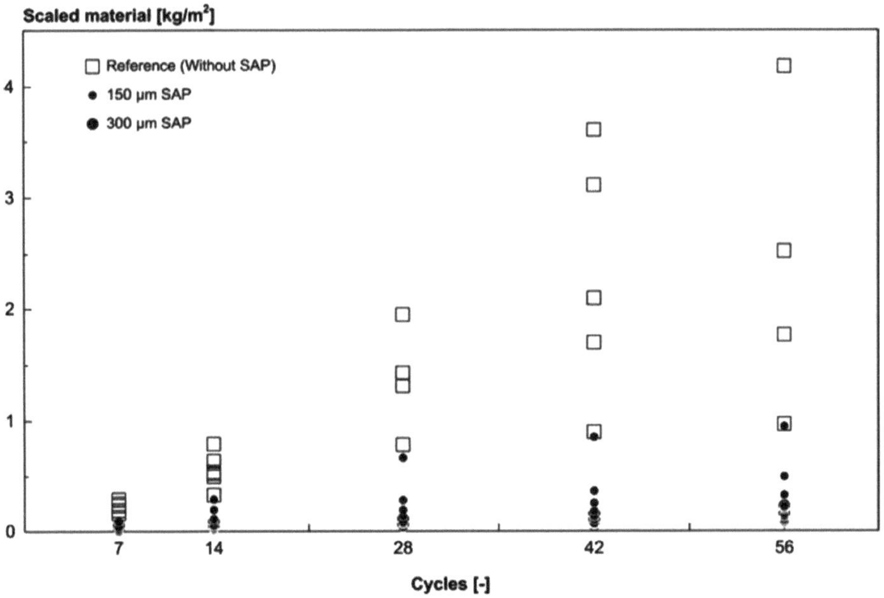

Fig. 9.11 Amount of scaled material as function of number of cycles [13]

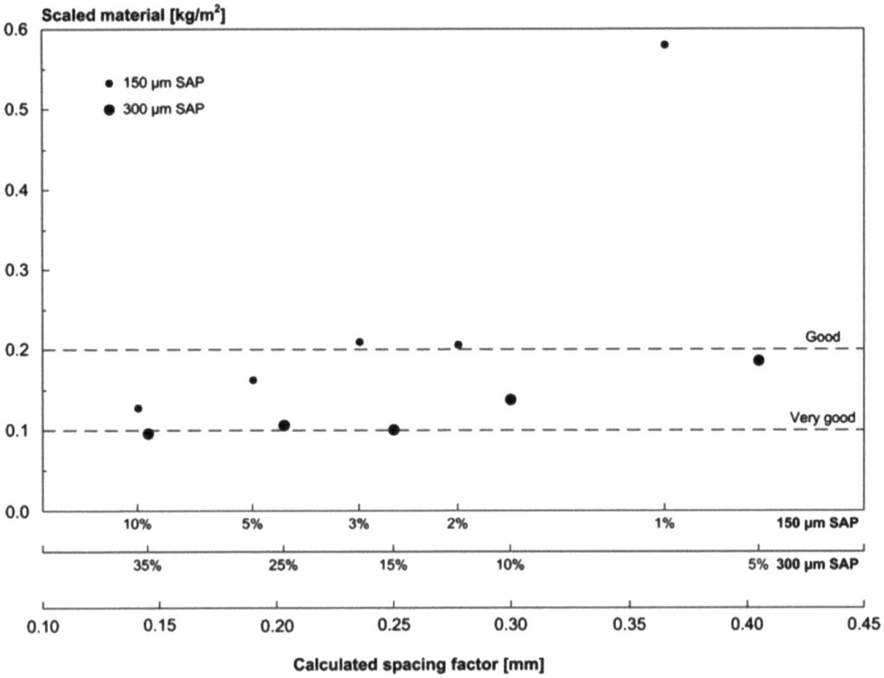

Fig. 9.12 Amount of scaling after 56 cycles as function of pore volume and spacing factor [13]

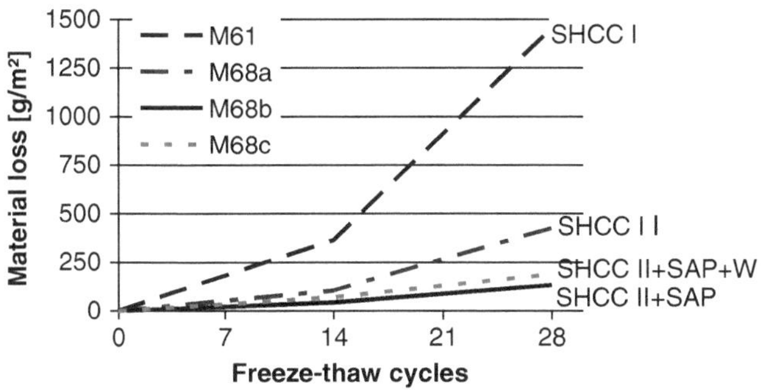

Fig. 9.13 Material loss in the CDF-test for different SHCC compositions [16]

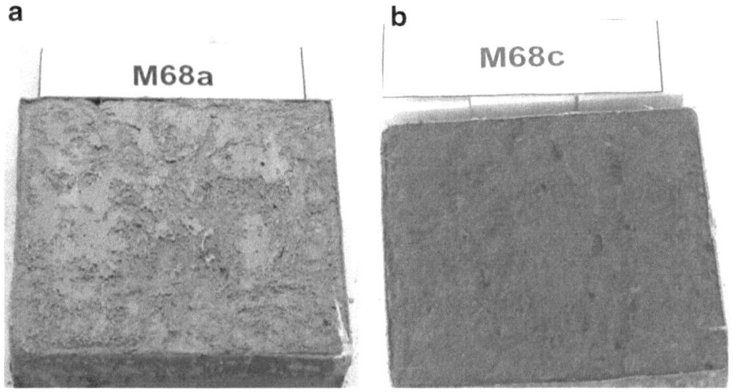

Fig. 9.14 Condition of the specimen surfaces after completion of the CDF-test: (**a**) SHCC without SAP, (**b**) SHCC with SAP and extra water [16]

the mix M68a contained no SAP or extra water, 2 kg SAP per cubic meter concrete were added to the mixes M68b and M68c. In the case of the mix M68c, apart from SAP some extra water was added for the purpose of internal curing. The water-to-binder ratio was 0.30 for M68a-c (without additional water), respectively. The fly ash to binder ratio was 0.55.

Figure 9.13 shows the results of the freeze-thaw investigation according to the CDF procedure. The material loss for the reference mixture M68a was about 500g/m². Due to the anchoring function of the fibres, not all the degraded material could be separated from the specimen during the cleaning of the test surface in an ultrasonic bath. Despite the relatively low material loss, the general appearance of the test surface was not satisfactory. The freeze-thaw resistance of specimens made of SHCC with the addition of SAP in respect of is considerably higher. Due to the well distributed system of SAP voids in mixes M68b and M68c, the loss in mass was clearly 130 g/m² and 190 g/m², respectively. The test surfaces appear similarly smooth for both these SHCC compositions, which lost only the very outer matrix layer (cf. Figure 9.14 for the mix M68c).

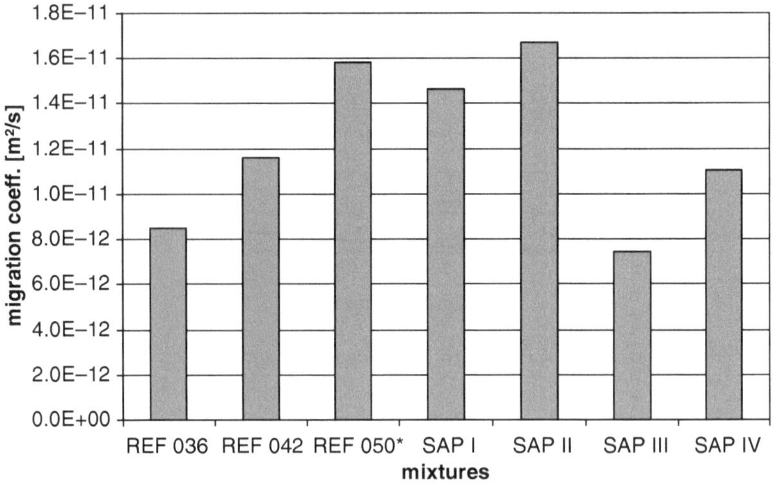

Fig. 9.15 Results of the chloride migration coefficient determination [18]

9.8 Chloride Migration

Chloride migration was determined with the rapid migration test described and developed by Tang [17]. The mixtures were identical to the mixtures used for the freeze-thaw tests of Table 9.3 The results of Figure 9.15 are within the range of values given by Gehlen [18] where for a CEM I 42.5 R cement type with a w/c-ratio of 0.40 a migration coefficient of $8.9 \cdot 10^{-12}$ m²/s was measured. SAP mixtures I to III had a w/c-ratio of 0.48 and the SAP IV mixture one of 0.52. After saturation of the SAP the resulting microstructure should correspond to the one of the REF 042 mixture. However, the SAP III and IV mixtures show smaller migration values comparable to the one of the REF 036 mixture. The other SAP mixtures had migration coefficients comparable to the one of the REF 050 mixture.

The results of the chloride migration tests showed large differences in the performance of the SAP mixtures. Therefore it seems not possible to deduce a final conclusion but there is a weak correlation between the median pore radius of the mixtures measured by MIP and the migration coefficients. A denser structure also results in smaller chloride migration. It seems unlikely that the SAPs produced an additional chemical effect influencing the measurement although they could theoretically bind ions. However, two reasons influenced the binding capacity of ions of SAPs in advance: First, the high ionic concentration of the pore fluid at the beginning of hydration. Second, the entanglement between hydration products and SAPs influenced the binding capacity. It is a result of hydration products growing into the SAP void.

References

[1] Reinhardt HW, Assmann A (2009) Enhanced durability of concrete by superabsorbant polymers. Proc. BMC 9, Warsaw, 291-300
[2] Jooss M, Reinhardt HW (2002) Permeability and diffusivity of concrete as function of temperature, Cement and Concrete Research 32:1497-1504
[3] Grube H (1983) Einfluss der Nachbehandlung auf die Porosität des Betons, Mitteilungen aus dem Forschungsinstitut des Vereins der österreichischen Zementfabrikanten, Bulletin No 36:54-59
[4] Jacobs FP (1994) Permeabilität und Porengefüge Zementgebundener Werkstoffe, PhD thesis ETH Zürich, Switzerland
[5] Graef H, Grube H (1986) Verfahren zur Prüfung der Durchlässigkeit von Mörtel und Beton gegenüber Gasen und Wasser, Betontechnische Berichte, Beton 36, No. 5:184-187 and No. 6:222-226
[6] Hall C, Hoff WS, Taylor SC, Wilson MA, Beom-Gi Yoon, Reinhardt HW, Sosoro M, Meredith P, Donald AM (1995) Water anomaly in capillary absorption by cement-based materials. Journal of Materials Science Letters 14:1178-1181
[7] Gaber K. (1989) Einfluss der Porengrößenverteilung in der Mörtelmatrix auf den Transport von Wasser, Chlorid und Sauerstoff im Beton, PhD thesis TU Darmstadt, Germany
[8] Reinhardt HW (2007) Beton-Kalender, WILEY-VCH, Berlin, 355-478
[9] RILEM Recommendation (1996) RILEM TC-117 FDC - CDF test – Test method for the freeze thaw resistance of concrete with sodium chloride solution, Materials & Structures 29:523-528
[10] Mönnig S, Lura P (2007) Superabsorbant polymers – An additive to increase the freeze-thaw resistance of high strength concrete. In C.U. Grosse (Ed.) Advances in Construction Materials, Springer Berlin, 351-358
[11] Powers TC (1949) The air requirement of frost resistant concrete. Proc. Highway Res. Board, No. 29
[12] Reinhardt HW, Assmann A, Ichimiya K (2009) Effect of SAP on durability of concrete. Meeting Rilem TC-SAP, Dresden, Germany
[13] Laustsen S, Hasholt MT, Jensen OM (2008) A new technology for air-entrainment of concrete. In Wei Sun, K. van Breugel, Changwen Miao, Guang Ye and Huisu Chen (Eds.) Microstructure Related Durability of Cementitious Composites, RILEM Publ. SARL, Bagneux, 1223-1230
[14] CEN/TS 12390-9 Testing hardened concrete - Part 9: Freeze-thaw resistance, August 2006
[15] SS 137244, Concrete testing – Hardened concrete – Scaling and freezing. Stockholm: Swedish Standards Institute, 2005, 10 pp
[16] Brüdern AE, Mechtcherine V (2010) Multifunctional use of SAP in Strain-hardening Cement-based Composites. International RILEM Conference on Use of Superabsorbent Polymers and Other New Additives in Concrete, Technical University of Denmark, Lyngby, RILEM Proceedings PRO 74, 11-22
[17] Tang L (1996) Chloride Transport in Concrete – Measurement and Prediction, PhD thesis, Chalmers University, Gothenburg, Sweden
[18] Gehlen C (2000) Probabilistische Lebensdauerbemessung von Stahlbetonbauwerken - Zuverlässigkeitsbetrachtungen zur wirksamen Vermeidung von Bewehrungskorrosion, DAfStb Bulletin No. 510, Berlin, Germany

Chapter 10
Practical Applications of Superabsorbent Polymers in Concrete and Other Building Materials

Daniel Cusson, Viktor Mechtcherine, and Pietro Lura

Abstract: Superabsorbent polymers (SAP) possess a number of features that make them attractive for use in many different applications. The aim of this chapter is to present an overview of existing and foreseen opportunities for the use of SAP in many different functions to improve the performance and durability of the built environment. Two case studies are also presented in this chapter: one on a thin-wall architectural structure in Germany, and another on shotcreting of wall panels in Denmark.

10.1 Introduction

The use of SAP in building materials is relatively recent, with most studies being conducted in the laboratory and very few known field applications. This chapter includes three major sections, which will discuss the improvement of concrete properties due to the use of SAP (section 10.2), potential applications for SAP in the construction sector (section 10.3), and case studies using SAP-modified concrete in the field (section 10.4).

D. Cusson (✉)
Institute for Research in Construction, National Research Council Canada, Canada
e-mail: daniel.cusson@nrc-cnrc.gc.ca

V. Mechtcherine
Institute of Construction Materials, Technische Universität Dresden, Germany

P. Lura
Empa, Materials Science and Technology, Switzerland

V. Mechtcherine and H.-W. Reinhardt (eds.), *Application of Superabsorbent Polymers (SAP) in Concrete Construction*, RILEM State of the Art Reports 2,
DOI 10.1007/978-94-007-2733-5_10, © RILEM 2012

10.2 Improvement of Properties

10.2.1 Shrinkage reduction

Controlling early-age cracking due to volume changes in concrete structures is essential to obtain long-term durability. Today, many concrete structures experience early-age shrinkage cracking. This is more common when high cement contents and low water-cement ratios are used to make the concrete, leading to autogenous shrinkage induced by self-desiccation. As presented in Chapter 7, SAP can be effectively used as a water-entraining agent in high-performance concrete to provide internal curing water that is needed to maximize cement hydration and minimize self-desiccation, with negligible adverse effects on other engineering properties.

10.2.2 Frost resistance

The production of concrete that is resistant to freezing and thawing requires special attention to some specific material parameters, including the air-void system, of which effectiveness is controlled by the volumetric air content, spacing and size of the air voids. To this effect, SAP particles can be engineered to provide an adequate pore system, since SAP particles can unswell during cement hydration and leave gas-filled voids, according to Jensen [1]. In fact, this concept has been recently demonstrated in the lab by Reinhardt et al. [2] and Laustsen et al. [3], where mixtures containing specific types of SAP were found to provide increased resistance to freezing and thawing in the presence of de-icing chemicals. It was hypothesized in [2] that SAP may have interacted with the superplasticizer to increase the air content, or very small air bubbles adhered to the SAP particles during mixing. In [3], the authors demonstrated the advantages of SAP-based technology over traditional chemical air entrainment, which are: stability of the air void system and improved control of both the amount of added air and the air void size. Their results clearly showed that the amount of scaled material depended solely on the amount of air voids in the concrete created by SAP, whereas the spacing factor was found to have only a minor influence.

As reported in [3], air entrainment with conventional air-entraining admixtures often encounters technical difficulties such as coalescence of air bubbles in fresh concrete, loss of air during consolidation or pumping, and chemical incompatibility with superplasticizers. Again, the use of SAP as an air-entraining agent could solve some of these difficulties in practice, since this technology is uninfluenced by the pumping and placing procedures [1].

More information on the effect of SAP addition on the frost resistance of concrete can be found in Chapter 9.

10.2.3 Rheology modification

The addition of SAP during concrete mixing can produce a considerable change in rheology, as observed by Jensen and Hansen [4]. The addition of SAP can allow a decrease in the free water-cement ratio, which can lead to an increase in both the yield stress and plastic viscosity, as mentioned in Chapter 5.

10.2.4 Controlled release

SAP may also be used to control the release of substances other than water that are dissolved in the SAP particles. Substances that are initially at a higher activity in the polymer will diffuse out of the particles into the surroundings, according to Buchholz and Graham [5]. Compared to other absorbent polymers, superabsorbent polymers have a special feature: their swelling depends on the pH of the swelling medium [5], which is a feature that may be used as switches for controlled release. Current commercial uses of SAP in this field are for pesticides, fertilizers and pharmaceuticals. As suggested in [1], a possible use of SAP as a controlled release agent in concrete could be for particular plasticizing admixtures that are more effective if they are first released shortly after initial contact between water and cement, at which time the pH of fresh concrete is relatively high.

10.2.5 Waterproofing

The volume increase of the gel of water-saturated swollen SAP can be used to form a barrier to water flow [5]. Sealing composites made by blending modified SAP into rubber (Tsubakimoto et al. [6]), or a thermoplastic elastomer [7] have been developed for sealing around the joints of various building materials. The composite may be used like mortar in joints and, if any gaps were left during construction or created after construction due to settlement, the SAP swells when in contact with leaking water and seals the joints, as suggested by Shimomura and Namba [8]. According to [5], such a composite was used in the construction of the Channel Tunnel between France and England.

10.2.6 Crack healing

Tsuji et al. [9, 10, 11] used SAP for blocking cracks in concrete. A special type of SAP that can hardly absorb alkaline water in fresh and hardened concrete was used to absorb neutral or acidic water infiltrating through cracks. In this case, SAP particles remain dormant (unswollen) within the concrete until a crack exposes them to the surface and water flowing through the crack causes them to swell. The effectiveness of the SAP was confirmed by the reduced permeability measured in the healed concrete.

A similar concept was proposed by Song et al. [12], in which a precursor solution of acrylic acid-co-acrylamide was injected into the concrete cracks together with an initiator and a cross-linker. The precursor was then activated with infrared radiation to initiate copolymerization. Preliminary tests on concrete cracks filled with large SAP particles (0.63-1.25 mm) showed reduced permeability of the repaired concrete. The swelling ratios of SAP in water, acidic, saline and basic solutions were also measured before and after accelerated ageing by ultraviolet radiation. In another study, Song et al. [13] used in-situ polymerization of SAP as a concrete surface treatment to improve sulphate resistance.

10.2.7 Surface curing

Poor surface curing conditions can reduce the durability of concrete surfaces due to high evaporation rates of water leading to plastic shrinkage cracking and slower development of surface strength. This could be prevented by applying a water-laden gel sheet to the concrete surface during the curing period, which would provide water to the surface, as required [5]. The gel layer is strengthened and protected from evaporation by applying a latex rubber coating on top of the gel [14].

Harrison [15] illustrated the use of a type of controlled-permeability formwork made of conventional formwork lined with SAP sheeting, which is impermeable to air and could absorb up to 200 times its weight of water. The SAP sheets are simply cut to length, folded over the form edge, and stapled to the form. During the compaction of concrete, some of the mix water escapes through the SAP form leaving the concrete in the cover zone with a reduced water-cement ratio. As a result, controlled-permeability formwork can achieve significant increase in concrete durability in the critical cover zone, improved surface appearance, and a substantial reduction in formwork pressures.

10.2.8 Fire protection

Jin et al. [16] used a SAP gel pre-soaked in an aqueous solution of calcium chloride to provide fire protection to building materials. This aqueous solution of calcium chloride has the ability to absorb water vapour or release it to the atmosphere until it reaches equilibrium with its surrounding. This approach was developed further by Asako et al. [17] who used SAP pre-saturated with a calcium chloride solution, mixed with cement, perlite, and water to obtain a fire-resistant mortar.

10.2.9 Removal of concrete contaminants

In [18], a technique was described by which radioactive isotopes present in the pore solution of concrete and other porous materials can be removed. A wetting agent and a SAP gel with engineered nanoparticles are applied onto the contaminated surface from a remote location. The wetting agent causes the radioactive material to re-suspend in the pore water. The SAP gel then draws the radioactive-laden water out of the pores, while the engineered nanoparticles irreversibly capture the radioactive molecules. The dried gel is then vacuumed and recycled, leaving only a small amount of radioactive waste. It was claimed that a single application of gel can remove up to 90% of the radioactive elements.

10.3 Potential Applications

This section presents expected applications, in which SAP could be effectively used to improve the construction process, performance, and durability of the built environment. It is noted that some of the following applications have already been introduced in previous sections and chapters; however, it was decided to present an overview of these applications for the sake of completeness.

10.3.1 Shotcreting

The use of SAP to increase viscosity and decrease rebound of shotcrete was the subject of a 1991 patent application from Snashall [19], in which it is proposed to premix SAP with the aggregate and 10% to 15% of aggregate weight of water, followed by a 10-minute stand in the mixer, and finally add the cement and the rest of the mix water. Water absorption by SAP is expected to happen in only 10 minutes. According to a later patent application from Jensen and Hansen [20], SAP can be added (i) dry in the nozzle to reduce the viscosity of a wet mix, or (ii) pre-swollen or partially pre-swollen for internal curing purposes. In the first case, a very rapid absorption of the SAP is desired to obtain a reduced viscosity before the shotcrete hits the wall. The second case has no such special requirement.

10.3.2 Backfilling

A water-blocking construction filler that is composed of cement, SAP, and an asphalt emulsion has been developed by Moriyoshi et al. [21]. The components may be mixed on site to form a gelled solid serving as a backfill material. For example, it could be used in tunnel construction to fill the gap between the tunnel liner and the walls of the boring [5]. A main advantage of this SAP-modified backfill material is its high deformability [21] compared to conventional backfill materials (e.g., gravel or sand) that often fracture during ground movement.

10.3.3 Soil stabilization

SAP may be blended with cement and other materials to form a composite soil stabilizer [22]. The composite may be added directly to the wet soil to absorb and gel any water present and, at the same time, form a rigid surface upon which the foundation footings of a building can be placed, as suggested in [5].

10.3.4 Smart paints

The swelling character of SAP could be used for water-blocking in designed waterproofed paints, as suggested in [1]. The deposited SAP particles of a carefully selected size range would inhibit quick ingress of water in wood, for example, without influencing evaporation of water from saturated wood.

10.3.5 Sensors

The swelling ability of SAP gels, their mechanical modulus, and sensitivity to changes in water content, pH and ionic strength make SAP suitable for the development of sensors [5], which could be used for the structural health monitoring of smart infrastructure and buildings. A pressure-sensitive switch based on a polyelectrolyte hydrogel has been developed by Sawahata et al. [23], and works on the principle that an electrical potential can be induced in a soft hydrogel by (i) applying mechanical stress in one part of the gel, or (ii) a change in the pH of the gel caused by deformation. By attaching wires to the gel, the potential difference generates a signal, of which intensity depends on the magnitude of the mechanical deformation. Water-sensing devices that use the high conductivity of swollen polyelectrolyte gels as the detection switch have also been developed [24], based on the fact that dry polymers do not conduct electricity and swollen polymers do conduct electricity and complete the circuit. The magnitude of the gel conductivity indicates the degree of water absorption.

With regard to concrete applications, it may be thought that such SAP-based sensors could be eventually developed to monitor pH or salt concentration in concrete structures, which are key parameters influencing chloride-induced or carbonation-induced corrosion of the steel reinforcement in concrete structures.

10.3.6 Other applications in concrete construction

The following applications, summarized from [25], are grouped in this section as they relate to a given commercial SAP composed of methyl cellulose, which is a well known viscosity modifier.

Adhesives and grouts – Methocellulose ethers have the ability to thicken adhesives and grouts, while making them easier to mix and apply. They provide water retention properties, which can help improving the workability, open times, adhesion, and sliding resistance of various cement-based adhesives.

Jointing compounds and mortars – They can be used in jointing compounds, due to their ability to provide improved bonding strength and workability.

Plasters and fillers – They can be used in plasters and fillers that are cement-based, gypsum-based, or dispersion-based. They help make plasters and fillers easier to mix and apply, while improving their bond strength, workability and water retention.

Self-leveling floor compounds – They can improve adhesion strength of self-leveling compounds used for covering poured surfaces before flooring installation.

Extruded cement panels – They can control water retention during the extrusion process to enhance the strength of extruded panels, allowing for durable materials.

10.4 Case Studies

10.4.1 FIFA World Cup Pavilion, Germany

This pavilion was built for the 2006 FIFA World Cup in Kaiserslautern, which was one of the host cities. As reported by Mechtcherine et al. [26], it was designed as a filigree, thin-walled structure with very slender columns (minimum wall thickness of 20 mm) and no conventional reinforcement (Figs. 10.1 and 10.2).

In order to meet the rigorous design requirements (including reduced autogenous shrinkage, high durability, enhanced ductility, self-compaction, and high-quality surface), self-compacting fibre-reinforced high-performance concrete with internal curing was developed by Dudziak and Mechtcherine [27]. SAP made of covalently cross-linked acrylamide/acrylic acid copolymers were used for internal curing, and a polycarboxylate superplasticizer was used to ensure adequate self-compaction of the concrete.

The concrete had a free water-cement ratio of 0.24 and included CEM I 42.5 R HS cement and micro silica as binders, a blend of quartz powders and basalt sands, 6-mm steel fibers, and SAP with an average particle size of 200 μm. The content of SAP was adjusted to provide up to 45 kg/m^3 of internal curing water in the concrete. A number of different compositions were developed, optimized and compared in the laboratory for this project. Table 10.1 provides information on the compositions

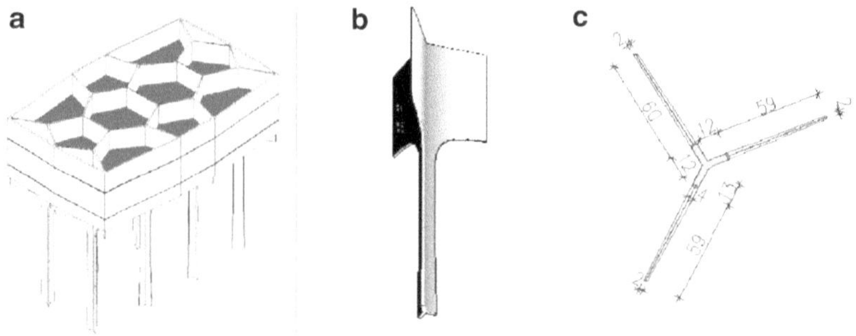

Fig. 10.1 FIFA World Cup Pavilion, City of Kaiserslautern, Germany [26] – (**a**) Schematic view of pavilion; (**b**) & (**c**) geometry of a column (dimensions in cm)

Fig. 10.2 Photograph of built pavilion [27]

of the UHPC mixture used for the construction of the pavilion (referred to as Pav-SAP) and of the corresponding reference mixture made without addition of SAP and extra water (referred to as Pav).

The laboratory tests conducted on the concrete revealed that the rheological properties of the fresh concrete (slump flow diameter of approximately 80 cm) were not adversely affected by the use of SAP and IC water when compared to a reference mixture made without SAP. The early-age autogenous shrinkage was greatly reduced by internal curing (from -605 $\mu\varepsilon$ to -72 $\mu\varepsilon$ at 7 days); however, the total shrinkage (including drying shrinkage) was only found to decrease slightly at a later age (from -1050 $\mu\varepsilon$ to -950 $\mu\varepsilon$ at 28 days). This finding indicates that the quantity of internal curing water was sufficient to prevent self-desiccation. Table 10.2 presents test results on the mechanical properties of the SAP-cured concretes.

Table 10.1 Compositions of UHPC with and without addition of SAP, adapted from [27]

Component	Pav	Pav-SAP
Cement, CEM I 42.5 R HS (kg/m³)	800	800
Silica fume (kg/m³)	120	120
Water, total (kg/m³)	179	203
SAP (% mass cement)	-	0.4
$(w/c)_{total}$ (incl. IC water)	0.25	0.28
$(w/c)_{effective} + (w/c)_{internal\ curing}$	0.25+0	0.25+0.03
Quartz powder (kg/m³)	206	195
Fine sand 0.125/0.5 mm (kg/m³)	229	217
Crushed basalt sand 0/2 mm (kg/m³)	184	173
Basalt split 2/5 mm (kg/m³)	522	493
Steel fibres 6 x 0.015 mm (kg/m³)	144	144
Superplasticiser (% mass cement)	4.3	4.3
Pigment Fe_2O_3 (kg/m³)	12	12

Table 10.2 Average mechanical properties of investigated concretes (standard deviations are given in parentheses), adapted from [27]

	Compressive strength [MPa] (tested on cubes)			Flexural strength [MPa] (tested on prisms)		
Mixture	Sealed 2d	Sealed 28d	Unsealed 28d	Sealed 2d	Sealed 28d	Unsealed 28d
Pav	96 (-)	139 (-)	140 (3)	13 (0.6)	15 (1.1)	21 (2.4)
Pav-SAP	85 (-)	131 (-)	140 (6)	15 (-)	19 (1.2)	16 (-)

(-) only two specimens tested

10.4.2 Shotcreting of Wall Panels, Lyngby, Denmark

According to Jensen [1], the thickening effect caused by the presence of SAP in concrete can be used advantageously in some practical situations such as pumping. Successful wet-mix shotcreting requires overcoming several technical challenges [1]. For instance, high slumps are usually required to achieve adequate pumpability, however, low slumps allow better thickness build-up and minimize rebound. Set-accelerating admixtures are often required but their use may lead to marked reductions in long-term compressive strengths, as found by Jolin & Beaupré [28]. Another difficulty is related to the control of air-entrainment in placed shotcrete.

It was tentatively shown in [1] that these difficulties can potentially be avoided by the dry addition of SAP in the nozzle during shotcreting (Fig. 10.3). The concrete had an initial w/c of 0.4 and contained 0.4% SAP with a water absorption near 15g of water per gram of dry SAP. It was observed that the uptake of water by SAP created a change in viscosity during placing and allowed the build-up of thick layers

Fig. 10.3 Shotcreting of wall panels with SAP-modified concrete, Lyngby, Denmark [1]

without the use of a set-accelerating admixture. In this case, SAP was added to shotcrete as a rheology modifier, however, other benefits may be found, such as internal water curing and mitigation of autogenous shrinkage (as explained in previous chapters).

10.5 Summary and Final Remarks

As shown in this chapter, superabsorbent polymers possess a large number of features that make them attractive for use in many different applications. Current applications using SAP in concrete structures have been reported and include: shrinkage reduction, frost protection, rheology modification, waterproofing, and fire protection, to name a few. Expected applications in the near future have also been identified, such as: shotcreting and backfilling, as well as potential uses in the development of innovative sensors.

It was shown through a case study that SAP-modified ultra-high strength concrete could be made in the field to build a thin-wall architectural structure that could meet the following rigorous design requirements: self-compaction, low autogenous shrinkage, enhanced ductility and high durability. A second case study provided evidence that challenges normally encountered in typical shotcreting applications could be overcome by the dry addition of SAP in the nozzle during the shotcreting operation.

There are clearly many opportunities to use SAP in many different functions to improve the performance and durability of the built environment. It is expected that future applications will increasingly move towards the use of SAP in the construction sector, as this new technology becomes better known through good practice and evidence of good performance records.

References

[1] Jensen OM (2008) Use of superabsorbent polymers in construction materials, 1st International Conference on Microstructure Related Durability of Cementitious Composites, Nanjing, China, 757–764

[2] Reinhardt HW, Assmann A, Mönnig S (2008) Superabsorbent polymers (SAP) – an admixture to increase the durability of concrete, 1st International Conference on Microstructure Related Durability of Cementitious Composites, Nanjing, China, 313–322

[3] Laustsen S, Hasholt MT, Jensen OM (2008) A new technology for air entrainment of concrete, 1st International Conference on Microstructure Related Durability of Cementitious Composites, Nanjing, China, 1223–1230

[4] Jensen OM, Hansen PF (2002) Water-entrained cement-based materials – II. Experimental observations, Cement and Concrete Research, 32(6):973–978

[5] Buchholz FL, Graham AT (1998) Modern Superabsorbent Polymer Technology, WILEY-VCH, New-York

[6] Tsubakimoto T, Shimomura T, Kobayashi H, (1987) Japan, Kokai Tokkyo Koho, 62–149, 335

[7] Suetsugu (1994) Japan, Kokai Tokkyo Koho, 06-157, 839

[8] Shimomura T, Namba T (1994) Superabosorbent Polymers, Science and Technology, Symposium Series 573, FL Buchholz, NA Peppas (eds), American Chemical Society, Washington DC, 112–115

[9] Tsuji M, Okuyama A, Enoki K, Suksawang S (1998) Development of new concrete admixture preventing from leakage of water through cracks, JCA Proc. of Cement & Concrete (Japan Cement Association) 52:418–423

[10] Tsuji M, Shitama K, Isobe D (1999) Basic Studies on Simplified Curing Technique, and Prevention of Initial Cracking and Leakage of Water through Cracks of Concrete by Applying Superabsorbent Polymers as New Concrete Admixture, Journal of the Society of Materials Science, Japan, 48(11):1308–1315

[11] Tsuji M, Koyano H, Okuyama A, Isobe D (1999) Study on method of test for leakage through cracks of hardened concrete, JCA Proc. of Cement & Concrete (Japan Cement Association) 53:462–468

[12] Song XF, Wei JF, He TSH (2009) A method to repair concrete leakage through cracks by synthesizing super-absorbent resin in situ, Construction and Building Materials 23(1):386–391

[13] Song XF, Wei JF, He TSH (2008) A novel method to improve sulfate resistance of concrete by surface treatment with super-absorbent resin synthesised in situ, Magazine of Concrete Research 60(1):49–55

[14] Onoda Cement Company Ltd, Japan, Kokai Tokkyo Koho, 62-56, 382, 1987

[15] Harrison T (1991) Introducing controlled permeability formwork. Increase concrete durability in the cover zone, Publication # C910198 Aberdeen's Concrete Construction 36(2):198–200

[16] Jin ZF, Asako Y, Yamaguchi Y, Yoshida H (2000) Thermal and water storage characteristics of super-absorbent polymer gel which absorbed aqueous, International Journal of Heat and Mass Transfer, 43(18):3407–3415
[17] Asako Y, Otaka T, Yamaguchi Y (2004) Fire resistance characteristics of materials with polymer gels which absorb aqueous solution of calcium chloride, Numerical Heat Transfer, Part A, 45:49–66
[18] Gel cleans radioactive concrete, Chemical Processing 67(9):9-11, 2004
[19] Snashall HT (1991) Cementitious mixes, South African Patent Application ZA9100876 A 19911224
[20] Jensen OM, Hansen PF (2001) Water-entrained cement-based materials, PCT Patent Application WO01/02317A1
[21] Moriyoshi A, Fukai I, Takeguchi M (1990) Nature, 344:230–232
[22] Nippon Synthetic Materials Ltd., Japan, Kokai Tokkyo Koho, 06-287, 556, 1994
[23] Sawahata K, Gong JP, Osada Y (1995) Macromol., Rapid Commun., 16:713–716
[24] Nippondenso Co. Ltd. (1994) Japan, Kokai Tokkyo Koho, 06-300, 724
[25] The Dow Chemical Co., Dow Wolff Cellulosics, Market and Applications - Construction Materials, http://www.dow.com/dowwolff/en/markets_apps/consmaterials.htm (accessed May 25, 2010)
[26] Mechtcherine V, Dudziak L, Schulze J, Staehr H (2006) Internal curing by super absorbent polymers (SAP) – Effects on material properties of self-compacting fibre-reinforced high performance concrete, Int RILEM Conf on Volume Changes of Hardening Concrete: Testing and Mitigation, Lyngby, Denmark, 87–96
[27] Dudziak L, Mechtcherine V 2008 Mitigation of volume changes of Ultra-High Performance Concrete (UHPC) by using Super Absorbent Polymers, 2nd Int Symp on Ultra High Performance Concrete, E Fehling et al (eds) Kassel University Press GmbH, 425–432
[28] Jolin M, Beaupré D (2003) Understanding wet-mix shotcrete: mix design, specification, and placement. Shotcrete, 6–12

Index

V. Mechtcherine and H.-W. Reinhardt (eds.), *Application of Superabsorbent Polymers (SAP) in Concrete Construction*, RILEM State of the Art Reports 2,
DOI 10.1007/978-94-007-2733-5, © RILEM 2012

RILEM Publications

The following list is presenting our global offer, sorted by series.

RILEM PROCEEDINGS

PRO 1: Durability of High Performance Concrete (1994) 266 pp., ISBN: 2-91214-303-9; e-ISBN: 2-35158-012-5; *Ed. H. Sommer*

PRO 2: Chloride Penetration into Concrete (1995) 496 pp., ISBN: 2-912143-00-4; e-ISBN: 2-912143-45-4; *Eds. L.-O. Nilsson and J.-P. Ollivier*

PRO 3: Evaluation and Strengthening of Existing Masonry Structures (1995) 234 pp., ISBN: 2-912143-02-0; e-ISBN: 2-351580-14-1; *Eds. L. Binda and C. Modena*

PRO 4: Concrete: From Material to Structure (1996) 360 pp., ISBN: 2-912143-04-7; e-ISBN: 2-35158-020-6; *Eds. J.-P. Bournazel and Y. Malier*

PRO 5: The Role of Admixtures in High Performance Concrete (1999) 520 pp., ISBN: 2-912143-05-5; e-ISBN: 2-35158-021-4; *Eds. J. G. Cabrera and R. Rivera-Villarreal*

PRO 6: High Performance Fiber Reinforced Cement Composites (HPFRCC 3) (1999) 686 pp., ISBN: 2-912143-06-3; e-ISBN: 2-35158-022-2; *Eds. H. W. Reinhardt and A. E. Naaman*

PRO 7: 1st International RILEM Symposium on Self-Compacting Concrete (1999) 804 pp., ISBN: 2-912143-09-8; e-ISBN: 2-912143-72-1; *Eds. Å. Skarendahl and Ö. Petersson*

PRO 8: International RILEM Symposium on Timber Engineering (1999) 860 pp., ISBN: 2-912143-10-1; e-ISBN: 2-35158-023-0; *Ed. L. Boström*

V. Mechtcherine and H.-W. Reinhardt (eds.), *Application of Superabsorbent Polymers (SAP) in Concrete Construction*, RILEM State of the Art Reports 2,
DOI 10.1007/978-94-007-2733-5, © RILEM 2012

PRO 9: 2nd International RILEM Symposium on Adhesion between Polymers and Concrete ISAP '99 (1999) 600 pp., ISBN: 2-912143-11-X; e-ISBN: 2-35158-024-9; *Eds. Y. Ohama and M. Puterman*

PRO 10: 3rd International RILEM Symposium on Durability of Building and Construction Sealants (2000) 360 pp., ISBN: 2-912143-13-6; e-ISBN: 2-351580-25-7; *Eds. A. T. Wolf*

PRO 11: 4th International RILEM Conference on Reflective Cracking in Pavements (2000) 549 pp., ISBN: 2-912143-14-4; e-ISBN: 2-35158-026-5; *Eds. A. O. Abd El Halim, D. A. Taylor and El H. H. Mohamed*

PRO 12: International RILEM Workshop on Historic Mortars: Characteristics and Tests (1999) 460 pp., ISBN: 2-912143-15-2; e-ISBN: 2-351580-27-3; *Eds. P. Bartos, C. Groot and J. J. Hughes*

PRO 13: 2nd International RILEM Symposium on Hydration and Setting (1997) 438 pp., ISBN: 2-912143-16-0; e-ISBN: 2-35158-028-1; *Ed. A. Nonat*

PRO 14: Integrated Life-Cycle Design of Materials and Structures (ILCDES 2000) (2000) 550 pp., ISBN: 951-758-408-3; e-ISBN: 2-351580-29-X, ISSN: 0356-9403; *Ed. S. Sarja*

PRO 15: Fifth RILEM Symposium on Fibre-Reinforced Concretes (FRC) – BEFIB'2000 (2000) 810 pp., ISBN: 2-912143-18-7; e-ISBN: 2-912143-73-X; *Eds. P. Rossi and G. Chanvillard*

PRO 16: Life Prediction and Management of Concrete Structures (2000) 242 pp., ISBN: 2-912143-19-5; e-ISBN: 2-351580-30-3; *Ed. D. Naus*

PRO 17: Shrinkage of Concrete – Shrinkage 2000 (2000) 586 pp., ISBN: 2-912143-20-9; e-ISBN: 2-351580-31-1; *Eds. V. Baroghel-Bouny and P.-C. Aïtcin*

PRO 18: Measurement and Interpretation of the On-Site Corrosion Rate (1999) 238 pp., ISBN: 2-912143-21-7; e-ISBN: 2-351580-32-X; *Eds. C. Andrade, C. Alonso, J. Fullea, J. Polimon and J. Rodriguez*

PRO 19: Testing and Modelling the Chloride Ingress into Concrete (2000) 516 pp., ISBN: 2-912143-22-5; e-ISBN: 2-351580-33-8; Soft cover, *Eds. C. Andrade and J. Kropp*

PRO 20: 1st International RILEM Workshop on Microbial Impacts on Building Materials (2000) 74 pp., e-ISBN: 2-35158-013-3; *Ed. M. Ribas Silva (CD 02)*

PRO 21: International RILEM Symposium on Connections between Steel and Concrete (2001) 1448 pp., ISBN: 2-912143-25-X; e-ISBN: 2-351580-34-6; *Ed. R. Eligehausen*

PRO 22: International RILEM Symposium on Joints in Timber Structures (2001) 672 pp., ISBN: 2-912143-28-4; e-ISBN: 2-351580-35-4; *Eds. S. Aicher and H.-W. Reinhardt*

PRO 23: International RILEM Conference on Early Age Cracking in Cementitious Systems (2003) 398 pp., ISBN: 2-912143-29-2; e-ISBN: 2-351580-36-2; *Eds. K. Kovler and A. Bentur*

PRO 24: 2nd International RILEM Workshop on Frost Resistance of Concrete (2002) 400 pp., ISBN: 2-912143-30-6; e-ISBN: 2-351580-37-0, Hard back; *Eds. M. J. Setzer, R. Auberg and H.-J. Keck*

PRO 25: International RILEM Workshop on Frost Damage in Concrete (1999) 312 pp., ISBN: 2-912143-31-4; e-ISBN: 2-351580-38-9, Soft cover; *Eds. D. J. Janssen, M. J. Setzer and M. B. Snyder*

PRO 26: International RILEM Workshop on On-Site Control and Evaluation of Masonry Structures (2003) 386 pp., ISBN: 2-912143-34-9; e-ISBN: 2-351580-14-1, Soft cover; *Eds. L. Binda and R. C. de Vekey*

PRO 27: International RILEM Symposium on Building Joint Sealants (1988) 240 pp., e-ISBN: 2-351580-15-X; *Ed. A. T. Wolf, (CD03)*

PRO 28: 6th International RILEM Symposium on Performance Testing and Evaluation of Bituminous Materials, PTEBM'03, Zurich, Switzerland (2003) 652 pp., ISBN: 2-912143-35-7; e-ISBN: 2-912143-77-2, Soft cover; *Ed. M. N. Partl (CD06)*

PRO 29: 2nd International RILEM Workshop on Life Prediction and Ageing Management of Concrete Structures, Paris, France (2003) 402 pp., ISBN: 2-912143-36-5; e-ISBN: 2-912143-78-0, Soft cover; *Ed. D. J. Naus*

PRO 30: 4th International RILEM Workshop on High Performance Fiber Reinforced Cement Composites – HPFRCC 4, University of Michigan, Ann Arbor, USA (2003) 562 pp., ISBN: 2-912143-37-3; e-ISBN: 2-912143-79-9, Hard back; *Eds. A. E. Naaman and H. W. Reinhardt*

PRO 31: International RILEM Workshop on Test and Design Methods for Steel Fibre Reinforced Concrete: Background and Experiences (2003) 230 pp., ISBN: 2-912143-38-1; e-ISBN: 2-351580-16-8, Soft cover; *Eds. B. Schnütgen and L. Vandewalle*

PRO 32: International Conference on Advances in Concrete and Structures, 2 volumes (2003) 1592 pp., ISBN (set): 2-912143-41-1; e-ISBN: 2-351580-17-6, Soft cover; *Eds. Ying-shu Yuan, Surendra P. Shah and Heng-lin Lü*

PRO 33: 3rd International Symposium on Self-Compacting Concrete (2003) 1048 pp., ISBN: 2-912143-42-X; e-ISBN: 2-912143-71-3, Soft cover; *Eds. Ó. Wallevik and I. Níelsson*

PRO 34: International RILEM Conference on Microbial Impact on Building Materials (2003) 108 pp., ISBN: 2-912143-43-8; e-ISBN: 2-351580-18-4; *Ed. M. Ribas Silva*

PRO 35: International RILEM TC 186-ISA on Internal Sulfate Attack and Delayed Ettringite Formation (2002) 316 pp., ISBN: 2-912143-44-6; e-ISBN: 2-912143-80-2, Soft cover; *Eds. K. Scrivener and J. Skalny*

PRO 36: International RILEM Symposium on Concrete Science and Engineering – A Tribute to Arnon Bentur (2004) 264 pp., ISBN: 2-912143-46-2; e-ISBN: 2-912143-58-6, Hard back; *Eds. K. Kovler, J. Marchand, S. Mindess and J. Weiss*

PRO 37: 5th International RILEM Conference on Cracking in Pavements – Mitigation, Risk Assessment and Prevention (2004) 740 pp., ISBN: 2-912143-47-0; e-ISBN: 2-912143-76-4, Hard back; *Eds. C. Petit, I. Al-Qadi and A. Millien*

PRO 38: 3rd International RILEM Workshop on Testing and Modelling the Chloride Ingress into Concrete (2002) 462 pp., ISBN: 2-912143-48-9; e-ISBN: 2-912143-57-8, Soft cover; *Eds. C. Andrade and J. Kropp*

PRO 39: 6th International RILEM Symposium on Fibre-Reinforced Concretes (BEFIB 2004), 2 volumes, (2004) 1536 pp., ISBN: 2-912143-51-9 (set); e-ISBN: 2-912143-74-8, Hard back; *Eds. M. Di Prisco, R. Felicetti and G. A. Plizzari*

PRO 40: International RILEM Conference on the Use of Recycled Materials in Buildings and Structures (2004) 1154 pp., ISBN: 2-912143-52-7 (set); e-ISBN: 2-912143-75-6, Soft cover; *Eds. E. Vázquez, Ch. F. Hendriks and G. M. T. Janssen*

PRO 41: RILEM International Symposium on Environment-Conscious Materials and Systems for Sustainable Development (2005) 450 pp., ISBN: 2-912143-55-1; e-ISBN: 2-912143-64-0, Soft cover; *Eds. N. Kashino and Y. Ohama*

PRO 42: SCC'2005 – China: 1st International Symposium on Design, Performance and Use of Self-Consolidating Concrete (2005) 726 pp., ISBN: 2-912143-61-6; e-ISBN: 2-912143-62-4, Hard back; *Eds. Zhiwu Yu, Caijun Shi, Kamal Henri Khayat and Youjun Xie*

PRO 43: International RILEM Workshop on Bonded Concrete Overlays (2004) 114 pp., e-ISBN: 2-912143-83-7; *Eds. J. L. Granju and J. Silfwerbrand*

PRO 44: 2nd International RILEM Workshop on Microbial Impacts on Building Materials (Brazil 2004) (CD11) 90 pp., e-ISBN: 2-912143-84-5; *Ed. M. Ribas Silva*

PRO 45: 2nd International Symposium on Nanotechnology in Construction, Bilbao, Spain (2005) 414 pp., ISBN: 2-912143-87-X; e-ISBN: 2-912143-88-8, Soft cover; *Eds. Peter J. M. Bartos, Yolanda de Miguel and Antonio Porro*

PRO 46: ConcreteLife'06 – International RILEM-JCI Seminar on Concrete Durability and Service Life Planning: Curing, Crack Control, Performance in Harsh Environments (2006) 526 pp., ISBN: 2-912143-89-6; e-ISBN: 2-912143-90-X, Hard back; *Ed. K. Kovler*

PRO 47: International RILEM Workshop on Performance Based Evaluation and Indicators for Concrete Durability (2007) 385 pp., ISBN: 978-2-912143-95-2; e-ISBN: 978-2-912143-96-9, Soft cover; *Eds. V. Baroghel-Bouny, C. Andrade, R. Torrent and K. Scrivener*

PRO 48: 1st International RILEM Symposium on Advances in Concrete through Science and Engineering (2004) 1616 pp., e-ISBN: 2-912143-92-6; *Eds. J. Weiss, K. Kovler, J. Marchand, and S. Mindess*

PRO 49: International RILEM Workshop on High Performance Fiber Reinforced Cementitious Composites in Structural Applications (2006) 598 pp., ISBN: 2-912143-93-4; e-ISBN: 2-912143-94-2, Soft cover; *Eds. G. Fischer and V.C. Li*

PRO 50: 1st International RILEM Symposium on Textile Reinforced Concrete (2006) 418 pp., ISBN: 2-912143-97-7; e-ISBN: 2-351580-08-7, Soft cover; *Eds. Josef Hegger, Wolfgang Brameshuber and Norbert Will*

PRO 51: 2nd International Symposium on Advances in Concrete through Science and Engineering (2006) 462 pp., ISBN: 2-35158-003-6; e-ISBN: 2-35158-002-8, Hard back; *Eds. J. Marchand, B. Bissonnette, R. Gagné, M. Jolin and F. Paradis*

PRO 52: Volume Changes of Hardening Concrete: Testing and Mitigation (2006) 428 pp., ISBN: 2-35158-004-4; e-ISBN: 2-35158-005-2, Soft cover; *Eds. O. M. Jensen, P. Lura and K. Kovler*

PRO 53: High Performance Fiber Reinforced Cement Composites HPFRCC5 (2007) 542 pp., ISBN: 978-2-35158-046-2; e-ISBN: 978-2-35158-089-9, Hard back; *Eds. H. W. Reinhardt and A. E. Naaman*

PRO 54: 5th International RILEM Symposium on Self-Compacting Concrete, 3 Volumes (2007) 1198 pp., ISBN: 978-2-35158-047-9; e-ISBN: 978-2-35158-088-2, Soft cover; *Eds. G. De Schutter and V. Boel*

PRO 55: International RILEM Symposium Photocatalysis, Environment and Construction Materials (2007) 350 pp., ISBN: 978-2-35158-056-1; e-ISBN: 978-2-35158-057-8, Soft cover; *Eds. P. Baglioni and L. Cassar*

PRO 56: International RILEM Workshop on Integral Service Life Modelling of Concrete Structures (2007) 458 pp., ISBN 978-2-35158-058-5; e-ISBN: 978-2-35158-090-5, Hard back; *Eds. R. M. Ferreira, J. Gulikers and C. Andrade*

PRO 57: RILEM Workshop on Performance of cement-based materials in aggressive aqueous environments (2008) 132 pp., e-ISBN: 978-2-35158-059-2; *Ed. N. De Belie*

PRO 58: International RILEM Symposium on Concrete Modelling CONMOD'08 (2008) 847 pp., ISBN: 978-2-35158-060-8; e-ISBN: 978-2-35158-076-9, Soft cover; *Eds. E. Schlangen and G. De Schutter*

PRO 59: International RILEM Conference on On Site Assessment of Concrete, Masonry and Timber Structures SACoMaTiS 2008, 2 volumes (2008) 1232 pp., ISBN: 978-2-35158-061-5 (set); e-ISBN: 978-2-35158-075-2, Hard back; *Eds. L. Binda, M. di Prisco and R. Felicetti*

PRO 60: Seventh RILEM International Symposium (BEFIB 2008) on Fibre Reinforced Concrete: Design and Applications (2008) 1181 pp, ISBN: 978-2-35158-064-6; e-ISBN: 978-2-35158-086-8, Hard back; *Ed. R. Gettu*

PRO 61: 1st International Conference on Microstructure Related Durability of Cementitious Composites (Nanjing), 2 volumes, (2008) 1524 pp., ISBN: 978-2-35158-065-3; e-ISBN: 978-2-35158-084-4; *Eds. W. Sun, K. van Breugel, C. Miao, G. Ye and H. Chen*

PRO 62: NSF/ RILEM Workshop: In-situ Evaluation of Historic Wood and Masonry Structures (2008) 130 pp., e-ISBN: 978-2-35158-068-4; *Eds. B. Kasal, R. Anthony and M. Drdácký*

PRO 63: Concrete in Aggressive Aqueous Environments: Performance, Testing and Modelling, 2 volumes, (2009) 631 pp., ISBN: 978-2-35158-071-4; e-ISBN: 978-2-35158-082-0, Soft cover; *Eds. M. G. Alexander and A. Bertron*

PRO 64: Long Term Performance of Cementitious Barriers and Reinforced Concrete in Nuclear Power Plants and Waste Management – NUCPERF 2009 (2009) 359 pp., ISBN: 978-2-35158-072-1; e-ISBN: 978-2-35158-087-5; *Eds. V. L'Hostis, R. Gens, C. Gallé*

PRO 65: Design Performance and Use of Self-consolidating Concrete, SCC'2009, (2009) 913 pp., ISBN: 978-2-35158-073-8; e-ISBN: 978-2-35158-093-6; *Eds. C. Shi, Z. Yu, K. H. Khayat and P. Yan*

PRO 66: Concrete Durability and Service Life Planning, 2nd International RILEM Workshop, ConcreteLife'09, (2009) 626 pp., ISBN: 978-2-35158-074-5; e-ISBN: 978-2-35158-085-1; *Ed. K. Kovler*

PRO 67: Repairs Mortars for Historic Masonry (2009) 397 pp., e-ISBN: 978-2-35158-083-7; *Ed. C. Groot*

PRO 68: Proceedings of the 3rd International RILEM Symposium on 'Rheology of Cement Suspensions such as Fresh Concrete' (2009) 372 pp., ISBN: 978-2-35158-091-2; e-ISBN: 978-2-35158-092-9; *Eds. O. H. Wallevik, S. Kubens and S. Oesterheld*

PRO 69: 3rd International PhD Student Workshop on 'Modelling the Durability of Reinforced Concrete' (2009) 122 pp., ISBN: 978-2-35158-095-0; e-ISBN: 978-2-35158-094-3; *Eds. R. M. Ferreira, J. Gulikers and C. Andrade*

PRO 71: Advances in Civil Engineering Materials, Proceedings of the 'The 50-year Teaching Anniversary of Prof. Sun Wei', (2010) 307 pp., ISBN: 978-2-35158-098-1; e-ISBN: 978-2-35158-099-8; *Eds. C. Miao, G. Ye, and H. Chen*

PRO 74: International RILEM Conference on 'Use of Superabsorsorbent Polymers and Other New Additive in Concrete' (2010) 374 pp., ISBN: 978-2-35158-104-9; e-ISBN: 978-2-35158-105-6; *Eds. O.M. Jensen, M.T. Hasholt, and S. Laustsen*

PRO 75: International Conference on 'Material Science - 2nd ICTRC - Textile Reinforced Concrete - Theme 1' (2010) 436 pp., ISBN: 978-2-35158-106-3; e-ISBN: 978-2-35158-107-0; *Ed. W. Brameshuber*

PRO 76: International Conference on 'Material Science - HetMat - Modelling of Heterogeneous Materials - Theme 2' (2010) 255 pp., ISBN: 978-2-35158-108-7; e-ISBN: 978-2-35158-109-4; *Ed. W. Brameshuber*

PRO 77: International Conference on 'Material Science - AdIPoC - Additions Improving Properties of Concrete - Theme 3' (2010) 459 pp., ISBN: 978-2-35158-110-0; e-ISBN: 978-2-35158-111-7; *Ed. W. Brameshuber*

PRO 78: 2nd Historic Mortars Conference and RILEM TC 203-RHM Final Workshop – HMC2010 (2010) 1416 pp., e-ISBN: 978-2-35158-112-4; *Eds J. Válek, C. Groot, and J. J. Hughes*

PRO 79: International RILEM Conference on Advances in Construction Materials Through Science and Engineering (2011) 213 pp., e-ISBN: 978-2-35158-117-9; *Eds Christopher Leung and K.T. Wan*

PRO 80: 2nd International RILEM Conference on Concrete Spalling due to Fire Exposure (2011) 453 pp., ISBN: 978-2-35158-118-6, e-ISBN: 978-2-35158-119-3; *Eds E.A.B. Koenders and F. Dehn*

RILEM REPORTS

Report 19: Considerations for Use in Managing the Aging of Nuclear Power Plant Concrete Structures (1999) 224 pp., ISBN: 2-912143-07-1; e-ISBN: 2-35158-039-7; *Ed. D. J. Naus*

Report 20: Engineering and Transport Properties of the Interfacial Transition Zone in Cementitious Composites (1999) 396 pp., ISBN: 2-912143-08-X; e-ISBN: 2-35158-040-0; *Eds. M. G. Alexander, G. Arliguie, G. Ballivy, A. Bentur and J. Marchand*

Report 21: Durability of Building Sealants (1999) 450 pp., ISBN: 2-912143-12-8; e-ISBN: 2-35158-041-9; *Ed. A. T. Wolf*

Report 22: Sustainable Raw Materials – Construction and Demolition Waste (2000) 202 pp., ISBN: 2-912143-17-9; e-ISBN: 2-35158-042-7; *Eds. C. F. Hendriks and H. S. Pietersen*

Report 23: Self-Compacting Concrete state-of-the-art report (2001) 166 pp., ISBN: 2-912143-23-3; e-ISBN: 2-912143-59-4, Soft cover; *Eds. Å. Skarendahl and Ö. Petersson*

Report 24: Workability and Rheology of Fresh Concrete: Compendium of Tests (2002) 154 pp., ISBN: 2-912143-32-2; e-ISBN: 2-35158-043-5, Soft cover; *Eds. P. J. M. Bartos, M. Sonebi and A. K. Tamimi*

Report 25: Early Age Cracking in Cementitious Systems (2003) 350 pp., ISBN: 2-912143-33-0; e-ISBN: 2-912143-63-2, Soft cover; *Ed. A. Bentur*

Report 26: Towards Sustainable Roofing (Joint Committee CIB/RILEM) (CD 07), (2001) 28 pp., e-ISBN: 2-912143-65-9; *Eds. Thomas W. Hutchinson and Keith Roberts*

Report 27: Condition Assessment of Roofs (Joint Committee CIB/RILEM) (CD 08), (2003) 12 pp., e-ISBN: 2-912143-66-7

Report 28: Final report of RILEM TC 167-COM 'Characterisation of Old Mortars with Respect to Their Repair' (2007) 192 pp., ISBN: 978-2-912143-56-3; e-ISBN: 978-2-912143-67-9, Soft cover; *Eds. C. Groot, G. Ashall and J. Hughes*

Report 29: Pavement Performance Prediction and Evaluation (PPPE): Interlaboratory Tests (2005) 194 pp., e-ISBN: 2-912143-68-3; *Eds. M. Partl and H. Piber*

Report 30: Final Report of RILEM TC 198-URM 'Use of Recycled Materials' (2005) 74 pp., ISBN: 2-912143-82-9; e-ISBN: 2-912143-69-1 – Soft cover; *Eds. Ch. F. Hendriks, G. M. T. Janssen and E. Vázquez*

Report 31: Final Report of RILEM TC 185-ATC 'Advanced testing of cement-based materials during setting and hardening' (2005) 362 pp., ISBN: 2-912143-81-0; e-ISBN: 2-912143-70-5 – Soft cover; *Eds. H. W. Reinhardt and C. U. Grosse*

Report 32: Probabilistic Assessment of Existing Structures. A JCSS publication (2001) 176 pp., ISBN 2-912143-24-1; e-ISBN: 2-912143-60-8 – Hard back; *Ed. D. Diamantidis*

Report 33: State-of-the-Art Report of RILEM Technical Committee TC 184-IFE 'Industrial Floors' (2006) 158 pp., ISBN 2-35158-006-0; e-ISBN: 2-35158-007-9, Soft cover; *Ed. P. Seidler*

Report 34: Report of RILEM Technical Committee TC 147-FMB 'Fracture mechanics applications to anchorage and bond' Tension of Reinforced Concrete Prisms – Round Robin Analysis and Tests on Bond (2001) 248 pp., e-ISBN 2-912143-91-8; *Eds. L. Elfgren and K. Noghabai*

Report 35: Final Report of RILEM Technical Committee TC 188-CSC 'Casting of Self Compacting Concrete' (2006) 40 pp., ISBN 2-35158-001-X; e-ISBN: 2-912143-98-5 – Soft cover; *Eds. Å. Skarendahl and P. Billberg*

Report 36: State-of-the-Art Report of RILEM Technical Committee TC 201-TRC 'Textile Reinforced Concrete' (2006) 292 pp., ISBN 2-912143-99-3; e-ISBN: 2-35158-000-1, Soft cover; *Ed. W. Brameshuber*

Report 37: State-of-the-Art Report of RILEM Technical Committee TC 192-ECM 'Environment-conscious construction materials and systems' (2007) 88 pp., ISBN: 978-2-35158-053-0; e-ISBN: 2-35158-079-0, Soft cover; *Eds. N. Kashino, D. Van Gemert and K. Imamoto*

Report 38: State-of-the-Art Report of RILEM Technical Committee TC 205-DSC 'Durability of Self-Compacting Concrete' (2007) 204 pp., ISBN: 978-2-35158-048-6; e-ISBN: 2-35158-077-6, Soft cover; *Eds. G. De Schutter and K. Audenaert*

Report 39: Final Report of RILEM Technical Committee TC 187-SOC 'Experimental determination of the stress-crack opening curve for concrete in tension' (2007) 54 pp., ISBN 978-2-35158-049-3; e-ISBN: 978-2-35158-078-3, Soft cover; *Ed. J. Planas*

Report 40: State-of-the-Art Report of RILEM Technical Committee TC 189-NEC 'Non-Destructive Evaluation of the Penetrability and Thickness of the Concrete Cover' (2007) 246 pp., ISBN 978-2-35158-054-7; e-ISBN: 978-2-35158-080-6, Soft cover; *Eds. R. Torrent and L. Fernández Luco*

Report 41: State-of-the-Art Report of RILEM Technical Committee TC 196-ICC 'Internal Curing of Concrete' (2007) 164 pp., ISBN: 978-2-35158-009-7; e-ISBN: 978-2-35158-082-0, Soft cover; *Eds. K. Kovler and O. M. Jensen*

Report 42: 'Acoustic Emission and Related Non-destructive Evaluation Techniques for Crack Detection and Damage Evaluation in Concrete' – Final Report of RILEM Technical Committee 212-ACD (2010) 12 pp., e-ISBN: 978-2-35158-100-1; *Ed. M. Ohtsu*

RILEM Publications published by Springer

RILEM BOOKSERIES (Proceedings)

VOL. 1: Design, Production and Placement of Self-Consolidating Concrete (2010) 466 pp., ISBN: 978-90-481-9663-0; e-ISBN: 978-90-481-9664-7, Hardcover; *Ed. K. Khayat and D. Feyes*

VOL. 2: High Performance Fiber Reinforced Cement Composites 6 - HPFRCC6 (2011) 584 pp., ISBN: 978-94-007-2435-8; e-ISBN: 978-94-007-2436-5, Hardcover; *Ed. G.J. Parra-Montesinos, H.W. Reinhardt and A.E. Naaman*

VOL. 5: Joint fib-RILEM Workshop on Modelling of Corroding Concrete Structures (2011) 290 pp., ISBN: 978-94-007-0676-7; e-ISBN: 978-94-007-0677-4, Hardcover; *Ed. C. Andrade and G. Mancini*

For the latest publications in the RILEM Bookseries, please visit http://www.springer.com/series/8781

RILEM STATE-OF-THE-ART REPORTS

VOL. 3: State-of-the-Art Report of RILEM Technical Committee TC 193-RLS 'Bonded Cement-Based Material Overlays for the Repair, the Lining or the Strengthening of Slabs or Pavements' (2011) 198 pp., ISBN: 978-94-007-1238-6; e-ISBN: 978-94-007-1239-3, Hardcover; *Ed. B. Bissonnette, L. Courard, D.W. Fowler and J-L. Granju*

VOL. 4: State-of-the-Art Report prepared by Subcommittee 2 of RILEM Technical Committee TC 208-HFC 'Durability of Strain-Hardening Fibre-Reinforced Cement-Based Composites' (SHCC) (2011) 151 pp., ISBN: 978-94-007-0337-7; e-ISBN: 978-94-007-0338-4, Hardcover; *Ed. G.P.A.G. van Zijl and F.H. Wittmann*

V. Mechtcherine and H.-W. Reinhardt (eds.), *Application of Superabsorbent Polymers (SAP) in Concrete Construction*, RILEM State of the Art Reports 2,
DOI 10.1007/978-94-007-2733-5, © RILEM 2012

VOL. 5: State-of-the-Art Report of RILEM Technical Committee TC 194-TDP 'Application of Titanium Dioxide Photocatalysis to Construction Materials' (2011) 60 pp., ISBN: 978-94-007-1296-6; e-ISBN: 978-94-007-1297-3, Hardcover; *Ed. Yoshihiko Ohama and Dionys Van Gemert*

VOL. 7: State-of-the-Art Report of RILEM Technical Committee TC 215-AST 'In Situ Assessment of Structural Timber' (2010) 152 pp., ISBN: 978-94-007-0559-3; e-ISBN: 978-94-007-0560-9, Hardcover; *Ed. B. Kasal and T. Tannert*

For the latest publications in the RILEM State-of-the-Art Reports, please visit http://www.springer.com/series/8780

MIX
Papier aus verantwortungsvollen Quellen
Paper from responsible sources
FSC® C105338

If you have any concerns about our products,
you can contact us on
ProductSafety@springernature.com

In case Publisher is established outside the EU,
the EU authorized representative is:
Springer Nature Customer Service Center GmbH
Europaplatz 3, 69115 Heidelberg, Germany

Printed by Libri Plureos GmbH
in Hamburg, Germany